# Python Experiments in Physics and Astronomy

*Python Experiments in Physics and Astronomy* acts as a resource for science and engineering students or faculty who would like to see how a diverse selection of topics can be analyzed and simulated using Python programs. The book also provides Python solutions that can be learned from and modified as needed. The book is mainly aimed at undergraduates, but since many science students and faculty have limited exposure to scientific programming, having a collection of examples that address curve-fitting, Fast Fourier Transforms, image photometry and image alignment, and many others could be very helpful not just for learning from, but also to support classroom projects and demonstrations.

**Key Features**:

- Features tutorials using Python for non-computer science students and faculty involved with scientific programming.

- Contains complete scientific programming examples for teaching and academic projects.

- Presents detailed Python solutions for Physics and Astronomy topics, not normally covered in depth, because they would be too time-consuming.

**Padraig Houlahan** recently retired from a career spanning both college teaching and IT management. He was the IT Director for Lowell Observatory for almost 17 years and a systems analyst for Oregon State University before that. Prior to working in IT, Houlahan was an Assistant Professor at Embry-Riddle Aeronautical University, and after a career in IT, he returned to teaching Physics for ERAU.

# Python Experiments in Physics and Astronomy

Padraig Houlahan

CRC Press
Taylor & Francis Group
Boca Raton  London  New York

CRC Press is an imprint of the
Taylor & Francis Group, an **informa** business

Designed cover image: Padraig Houlahan

First edition published 2025
by CRC Press
2385 NW Executive Center Drive, Suite 320, Boca Raton FL 33431

and by CRC Press
4 Park Square, Milton Park, Abingdon, Oxon, OX14 4RN

*CRC Press is an imprint of Taylor & Francis Group, LLC*

© 2025 Padraig Houlahan

ISBN: 978-1-032-98189-5 (hbk)
ISBN: 978-1-032-98699-9 (pbk)
ISBN: 978-1-003-60004-6 (ebk)

DOI: 10.1201/9781003600046

Typeset in Minion
by SPi Technologies India Pvt Ltd (Straive)

# Contents

# Preface

THIS WORK IS A result of a desire, after I retired, to build a collection of programming projects concerning topics I had encountered during my college years, both as a student and teacher. The motivations were many. In some cases, I wanted to see how some complex or abstract technical problems were solved, such as modelling how hydrogen spectral lines changed with temperature or how images of star fields could be aligned automatically. In others, it was to explore real-world systems such as projectile motion where the standard equations of motion breakdown at high velocities because of drag or to see for myself how adding additional mass (Dark Matter) to galaxies can account for observed rotation speeds.

Creating and designing models and simulations of phenomena in physics and astronomy is not only a great way to deeply learn how the systems work through implementing their underlying equations, but it is also a terrific way to share knowledge with others. Like others I think of my own experiences when either as a graduate student or researcher or a lecturer, there were many times I wished I had decent and complete code examples to refer to, to see how solutions were implemented; I am unashamedly a 'monkey see, monkey do' kind of learner. I also feel that whether it is for boot-strapping a project or to support teaching presentations, such simulations can play a vital role, allowing lecturers to assign projects where students and faculty with limited programming experience and time can take an existing simulation and modify it to suit; perhaps to explore a whole new area by working with a different data column in a catalog or even a different catalog altogether.

Obviously, there are limitations in what can be achieved in a single volume. I have emphasized making code immediately available to the reader, and to save space, I limit the normally expected in-code comments and self-documentation and cut corners by not exhaustively testing and hardening the code. My assumption is that those who would be interested in this work will have enough of a computing background, even if meager, and (most importantly) a willingness, to look at the code examples and figure out how they work – I avoid clever and abstract computing techniques as much as possible to encourage this reverse engineering. From my own experience, both as a researcher and a teacher, I understand there are times when researchers of all ages and faculty are at a fragile stage of learning where they understand the broad strokes of computing but face an uphill struggle to get over the next hump in the learning curve and can benefit from good examples. The examples included here are intended for them. They are not perfect, and they are deliberately primitive when needed. But they are functional. And fun!

# Introduction

THIS BOOK EXPLORES A whimsical collection of topics from astronomy and physics with a common underlying theme – when encountered in the classroom, they often elicit an 'How was that done?' or 'I wish I could see that demonstrated!' response. To this end, I address the topics by constructing software that models the systems, based on their underlying scientific concepts, because I believe relatively simple software models can be of great benefit to students and faculty wrestling with learning and teaching complex processes. Having access to the simulations' source code, let the user see for themselves how problems were solved, which allows them to add additional code to reveal more details of interest.

Most topics here are in the realm of the applied, such as the various astronomical image processing tasks we address, that is, photometry, image alignment, and extracting object data from images; or for example, modelling real-world artillery shell trajectories with ranges four or five times smaller than the drag free models encountered in introductory courses.

But even the applied can be very abstract. For example, a well-known challenge in modelling is the creation of sample sets drawn from a probability density function. How is this to be done? The answer of course is to sample the corresponding Cumulative Density Function. While the proof can be presented very succinctly, it can leave the student with a sense of wondering if they're missing something and wishing for tangible demonstrations. We address this by exploring a variety of density functions and showing how their CDFs can be created and then sampled, not just to verify the procedure but to also show how to proceed when unique or non-standard distributions must be addressed.

In a course on integral calculus, students usually learn how to calculate the gravitational force for a sphere of radius R, with the perhaps surprising result that the gravitational force at a distance $r < R$ from the center only depends on the mass inside of r; when r > R, it's as if all the sphere's mass is concentrated at the center. By modelling mass distributions and adding up the effects from all points in the distribution, we can demonstrate these effects and explore if there are similar results for other geometries, such as disks, cubes, and rings. The great thing about this kind of modelling is that once you figure how to create the distribution, you can find the gravitational effects at any location, not just along an axis of symmetry which simplifies the analytic approach.

In astronomy courses, students are taught to account for galaxy rotation curve behaviors, where the speeds of stars orbiting a spiral galaxy's core seem to level off with distance, instead of decreasing with distance. To resolve this, additional mass must be added to the galaxy outer regions – Dark Matter. Trying to analyze the gravity fields from mass distributions involving spheres, disks, and shells from arbitrary directions is not always easy and often will not result in elegant equations. Under these circumstances a brute force mechanism where geometrical distributions can be co-added and the effects of all particles added to determine the field at the sampling locations of interest is easier; a graph of the result can be very compelling for the student. Better yet, it is straightforward to switch from gravitational forces to electrostatic ones, and a whole slew of charge distributions can now be studied.

I also explore how hydrogen spectral lines are generated and create the famous Balmer spectral series. The underlying mechanisms and mathematics are complex – even for my primitive approach, but in showing how we can estimate how the line intensity varies with temperature, and how temperature influences which of the series will be the brightest, all the while considering the degree the gas (hydrogen) is becoming ionized, the reader can explore these effects for themselves and try other simple atoms perhaps. There really is something wonderful after undertaking an indepth mathematical study of a process, being able to model it, and for the Balmer series, not only do we do that, but we use our results to create simple synthetic spectra, which I think brings a sense of completeness to the effort, where the theoretical is contrasted with the observed graphically.

In assembling this diverse collection of topics, and in providing Python classes and sample codes to explore them, I hope the reader will be motivated to replicate some of the scenarios for themselves and learn from seeing the solutions presented. The demonstration codes are very basic and clean to make them as transparent and as brief as possible since I don't want the reader to feel like they are facing an intimidating wall of dense code.

My hope is the reader will appreciate being able to look at and study the coding solutions I provide here and borrow from them and build upon them for their purposes. I think there are many possible applications at the college level where faculty and students could build on the examples for end of course, or senior thesis projects; for projects which would be too complex unless they could be jump-started by using demonstrations like the ones here. Some of the examples here could serve as an independent approach to cross-check other solutions, and to focus on a core process, to see how it works. For example, there are freely available photometry software packages that can be used to measure a star's brightness. Some are quite overwhelming in their complexity and appear to suffer from too much building the new on top of the old, which introduces quirkiness and instability. I would argue that when first encountering photometry, a simple, transparent, application that could be easily explained, and modified as needed, would be a powerful resource for both the teacher and the student. At all times, I do keep in mind the benefits of giving the reader access to model code, and hope to encourage them to either modify or develop their own, since I strongly believe the process of developing and coding models helps foster a better understanding and appreciation of the underlying processes being investigated.

The demonstration Python code presented in this project was designed mostly for clarity and less so to be efficient. There is an assumption that the reader has sufficient background to be able to learn from the examples or is willing to work toward that level of proficiency. To reduce the code example sizes, they are not heavily commented, and key details are described in the Programming Notes sections. Note also the examples are often 'bare bones' and have not been developed to meet all possible applications; they are for demonstration purposes intended to illustrate an approach to solutions; hardening the code to be more robust would greatly increase the size. Regardless, the examples still serve as a first step for the reader to build upon.

Efficient code can be very cryptic to read and maintain and would defeat our teaching purpose. For example, when manipulating class parameters in a function, I will often do something like x = self.x, and y = self.y so I have local copies of the class variables, and then write a much easier equation like 'z = x+y' instead of 'z = self.x + self.y.' Often, I will, somewhat inefficiently extract data columns, such as for coordinates, from a dataframe just for the convenience of being able to use them directly in an equation instead of a more cumbersome dataframe indexing form.

It's all a matter of judgment and personal style. And for terminology, while there is a benefit in referring to a Python class function as a 'method,' I prefer to use the term 'function,' probably because I look at the problem from a mathematical perspective. And similarly for class attributes, I prefer to call them variables or parameters.

Since a major goal for this book is to allow the reader to see how things were done, this presents a dilemma in that while there is a need to show the code used and to explain it in sufficient detail so a reader with modest programming skills could understand what was being done, and this kind of detailed discussion about the mechanics and techniques can distract from the narrative. For this reason, I will regularly provide an overview of a chapter's underlying Python code before providing the complete listing, so it is available for reference.

## EQUATIONS AND UNITS

When teaching physics and astronomy courses, the student is introduced to many different fascinating and wonderful concepts that are generally summarized in the form of equations. Scientists like to use equations because once an equation is developed, it can be used to make predictions which can be compared with observations. If there's disagreement, then the scientist is alerted there are limits on how useful the equation is, and they can work toward improving it.

When teaching students about phenomena and their underlying equations, they will normally be taught how the equations are derived and be subjected to lots of homework to practice using them, which not only makes the equations more familiar and thereby less threatening but also prepares students for subsequent lessons with more equations!

We can approach using an equation in different ways. For example, a car moving from rest might obey the rule $\mathbf{v} = \mathbf{10t}$, where $\mathbf{v}$ is the speed after travelling for time $\mathbf{t}$ under the influence of acceleration $\mathbf{a=10}$. If I know $\mathbf{t}$, I can calculate $\mathbf{v}$, and if I know $\mathbf{v}$, I can solve for $\mathbf{t}$. But what if we had a more complicated relationship relating $\mathbf{t}$ and $\mathbf{v}$? What if, because of lumpy roads, variable wind resistance, engine sputter and ice patches on the road, and perhaps inebriation, the relationship was something like:

$$\mathbf{v} = \mathbf{9.1 + 3.9t - 0.4\,t^{2.23} + .01\sin\left(3\cos\left(2t\right) - 8t\right)?}$$

If $\mathbf{v}$ is 10, what is $\mathbf{t}$? Good luck solving that!

One way to solve this problem would be to create a table of **v** values, using a range of **t** values and then seeing which **t** best matches the observed **v**. To get better results, we might have to refine the granularity of the **t** range. We would in fact be conducting a numerical experiment, exploring **t** values to see how they match our **v** observations. To what end? First to get an answer as to what **t** value(s) resulted in a specific **v**, and second, to get a sense of our system – perhaps by plotting **v** vs **t**. The latter can be extremely critical to helping gain insight into how the system described by our equation behaves – one of the most important payoffs for engaging in numerical experiments.

Knowing velocity, we could derive the distance travelled by integrating it with respect to time, but in this case, the analytic integration would be tricky. On the other hand, undertaking the integration numerically, by simply adding up **v(t)*Δt** over a range of **t** values, where **Δt** is a time interval, would be a relatively straightforward computing problem, so a numerical solution can often be easier to attain than an analytical. Obviously, such an approach would be accessible to students without advanced calculus skills. It is also worth noting that we could find the distance travelled, from a set of **v(t)** observations using numerical integration, even if we did not have a good analytical equation to match the data.

Beyond trying to evaluate equations, computer-based numerical approaches allow us to study complicated systems, to easily adjust various parameters to better understand how the system works. Even for relatively simple systems with simple equations, such as Kepler's Law of Planetary Motion which states the square of the period is proportional to the cube of the orbit size ($\mathbf{P}^2 = \mathbf{k}\,\mathbf{a}^3$), where **k** is a constant, there is something wonderful in showing students how it works remarkably well using **P** and **a** values for the Solar System's planets. Better yet, it must also hold true for the dozens of moons that orbit Jupiter and Saturn, and this is easily demonstrated using a program or a spreadsheet. For the teacher or student wishing to explore this, we discuss how to do this in one of our later chapters. Theory is great, but it's comforting to see it works when applied to different systems; Kepler's law truly appears to be universal, and not just an answer that only works for the planets since it works for moons and satellites also – that is a payoff for doing that chapter's numerical experiments. In this sense, conducting such numerical experiments and comparing results with observations deepens our understanding and increases our confidence in their reliability and in our understanding of the system being studied.

In this book, we explore a collection of topics/systems from physics and astronomy by modelling the systems using software, which then makes it easy to modify and tweak our models and see how results compare with observation. This achieves many things: First, it can validate our model system's underlying equations and second, it can help deepen our understanding of how the equations work. Third, it prepares us to make improved models and to explore other similar kinds of systems, the results of which can help guide theoreticians in giving them outcomes against which their equations can be tested, but these equations might themselves be used to improve the models!

Most of the projects we will explore will often be very simplified, but the beauty of developing a model is it can often be improved upon by adding a new feature to the software – an essential aspect of all research. For our purposes, the terms numerical experiment and model will be treated the same.

Numerical experiments can be done using spreadsheets and computer programs. Spreadsheets can be very effective for simple systems, but with more complicated scenarios, or more demanding presentation expectations, the control offered by a decent programming software such as Python, to me, makes it the preferred approach.

In the previous discussion, I avoided using physical units such as meters, seconds, or miles-per-hour. Under most circumstances, this would and should be considered a bad practice, and I would hold my students accountable for not including correct units when doing homework. But there are times when we can assume they are implied and not explicitly written for the sake of clarity. It is assumed that correct units will be used when needed.

For our purposes, the choice of units is simply a choice setting the scales for a plot's axes. For example, Newton's Law of Gravity uses the masses of two objects, and their separation to calculate the force of attraction between them:

$$F = G\frac{mM}{r^2}$$

To set the correct scaling for the force to match observation, when using meters and kilograms, for distance and mass, respectively, a scaling constant $G = 6.6 \times 10^{-11}$ is used; if using centimeters and grams, G would be $6.6 \times 10^{-8}$. Is there a system of units where $G = 1$? Sure. I could say if m and

M are the same and I call that mass 1, and they are separated by a distance I will also call 1, then I now have a force F = 1 in that particular scheme. Of course, if I want to compare my results with normal usage, I will have to figure out how to convert them (inevitably using G). But here's the thing: No matter what system of units I use, for a given initial m, M, and r, if I make m or M ten times larger, F will be ten times larger; if I make r three times greater, F will become nine times smaller, in other words all systems will have their forces change by the same multiplier.

We are, in a sense, treating the equation as having two parts: There is the structural part with m, M, and r, which captures how gravity works, and there is the scaling part. The structural part will always have the same behavior, whether it's for the gravitational attraction of an electron orbiting a proton, or the Moon around the Earth, or the Sun orbiting the Galaxy: Doubling the separation will reduce the effect by a factor of four and so on – no matter what units are used. So, for clarity, and convenience, I will sometimes omit the scaling constant and choose convenient units if I'm mainly interested in seeing how a system behaves – mainly in seeing how its plots look.

## CODE EXAMPLES

The code examples were written in Python, and use popular supporting libraries such as Matplotlib, Pandas, Numpy, Math, Itertools, and Scikit. As presented, they can be run from an IDE such as Spyder, but any IDE should work. There is no reason they couldn't be saved as standalone applications, but this was not done here to avoid the headaches of saving code in a robust form that would run on different operating systems and also because the examples generally consist of a Python class developed for a particular task, and a small test application that uses the class but with hardwired parameters like filenames and number of input lines to read into the test code. Some mechanisms, such as a configuration file or command line argument reading capability, would need to be added.

I also chose to not write GUI/message-driven code in general. Elsewhere, I have specifically written on how this could be done, but for the most part, this keeps the focus on the project's science and not the complex distraction of managing GUI widgets and event-driven code. When dealing with a project like this, where showing the code solutions is a major part of hope for reader's benefit, a balance must be struck between how much of the code to explain and how much to show in a chapter; too much and the science narrative suffers, too little and the coded solutions and purpose are unclear. I have tried to strike a balance where in most cases, important code sections are included in chapter, but complete listings are provided at the chapter ends so the reader can quickly refer to the structure if they want.

In a work like this, the question arises as to how should code segments be shown. Stylistically, I use two methods to show code examples. I generally use a simple cut and paste from the IDE, and then use a terrific Microsoft Word Add In called 'Easy Code Formatter' which will build a text-based representation that can flow across page boundaries. A very nice feature is that code listings can be split by selecting/creating a blank line and setting the style to Normal for that line, which retains the formatted computer code look-and-feel, enclosed by the accenting border. (Since both methods make code examples available with their lines numbered, I will often simply refer to the line number, and omit the figure number, for brevity.) There is no one perfect solution, but these are useful options for readers who might need to produce their own documents.

Since a major goal for this book is to make complete code examples immediately available for the reader to peruse and study, I decided to include the major code examples in each chapter, normally at the end, for reference. While this can present the reader with a large block of code all at once, I felt that efforts to subdivide code and discuss code subdivisions one at a time actually made things more confusing. Simply referring to line numbers within the one code block was clearer and more manageable. This approach naturally led to adding reference sections in each chapter called Programming Notes that discussed some of the more important features in the code, without distracting for the chapter's primary narrative.

Finally, I will normally use bold-face style when referring to code elements (functions, classes, variables, equations), but leave mathematical equations and entities in normal.

## CHAPTER 1

# Python and Object-Oriented Design Notes

I N THIS BOOK, WE explore different topics using the Python programming language. We assume the reader has some ability and willingness to learn from our examples and has some familiarity with Python. But while it is beyond the scope of this book to explain Python's details, there are some coding and design topics that kept reappearing, mainly because there is a common thread of scientific programming and data running through all chapters. In this chapter, we will discuss some of these topics concerning Object-Oriented Design and also a selection of Python data manipulation tools useful for data array and numerical calculations that proved very useful.

## OOD NOTES

For most of the code examples presented in this work, we rely heavily on Python's Object-Oriented Design (OOD) techniques. While OOD can be quite nuanced when used to its full potential, it turns out even a little goes a long way, and for the kind of scientific programming used here, being able to use a few fundamental constructs allowed for the creation of code that was quite manageable, easy to organize, and easy to apply. In this chapter, we discuss some basic OOD concepts that were used repeatedly by the different projects covered.

DOI: 10.1201/9781003600046-1

Using an OOD approach is really the way to properly manage your code. You will likely find that once you start adding features and capabilities, you will quickly end up with hundreds of lines of code in a file, and it can become tedious to have to jump around when making small adjustments. OOD's class constructs will go a long way toward making your code manageable and re-usable, but even then, you might be faced with bulky files that are burdensome. One solution would be to add other classes or sub-classes to your design, but if you are happy with the current design, but simply want to de-bulk the file, a possible (but ugly) strategy would be to place class methods in separate files and to import them; however, if importing a collection of methods (defs) into a class, make sure the import statement is in the class and everything is properly indented!

For some of our numerical experiments, we would like to run particular models from a class of models. Since many models would have similarities, they should have methods in common that could be stored in the parent class. We would also like our models to be easily configurable, since we don't have the benefit of a dashboard with interfaces for sliders and buttons; this can be done by using command line arguments.

We will now illustrate some of these concepts by developing a system with a parent class and two subclasses. What we wish to demonstrate here are how child objects inherit properties from their parents, if configured to do so.

Figure 1.1 shows three classes, a parent, child1, and child2. Both child1 and child2 use the parent class. The parent class has a variable **b** set to 8, and two functions, **__init__**() and **pprint**(). Its **__init__**() function will

```
 1. class parent:
 2.     b = 8
 3.     def __init__(self,x):
 4.         self.a = 1
 5.         self.x = x
 6.         self.pprint('Hi - this is the parent!')
 7.
 8.     def pprint(self,text):
 9.         print(text)
10.
11.
12. class child1(parent):
13.     def __init__(self,x):
14.         self.x = x
15.         self.pprint('Hi - this is child1!')
16.
17. class child2(parent):
18.     def __init__(self,z):
19.         super().__init__(9)
20.         self.x = z
21.
```

FIGURE 1.1    Class definitions for parent, child1, and child2 example.

```
 1. p = parent(4)
 2. Hi - this is the parent!
 3.
 4. p.a
 5. Out[194]: 1
 6.
 7. p.b
 8. Out[195]: 8
 9.
10. p.x
11. Out[196]: 4
12.
```

FIGURE 1.2    Instantiating a parent object, p.

set the values of **self.a** and **self.x**, with **self.x** depending on the value passed as an argument.

However, variables **a** and **x** only come into existence when a parent object is created (i.e., 'the parent is instantiated') or if a child specifically invokes the parent's version of __**init**__().

If we run the code in Figure 1.1, nothing appears to happen, but our IDE will have learned about the classes. Now let's use the console to create a parent object by running the command '**p** = **parent(4)**' and test some of its variables (see Figure 1.2). We see that all three variables have been set. Because the __**init**__() function was automatically run during instantiation and invokes **pprint()**, the information string 'Hi – this is the parent!' was also printed.

Now let's see what happens when we create a child1 object from the console and test its variables (see Figure 1.3).

```
 1. c1 = child1()
 2. Traceback (most recent call last):
 3.
 4.   Cell In[12], line 1
 5.     c1 = child1()
 6.
 7. TypeError: child1.__init__() missing 1 required positional argument: 'x'
 8.
 9.
10. c1=child1(5)
11. Hi - this is child1!
12.
13. c1.a
14. Traceback (most recent call last):
15.
16.   Cell In[14], line 1
17.     c1.a
18.
19. AttributeError: 'child1' object has no attribute 'a'
20.
21.
22. c1.x
23. Out[15]: 5
24.
```

FIGURE 1.3    Testing the child1 class.

Our first attempt at creating object c1 failed (line 7), because its initialization (Figure 1.1, line 13) expected an input variable.

Retrying with an argument specified (line 10) worked, but when we test for variables, **c1.a** didn't exist, but **c1.x** did (see lines 13 and 22.) To understand what happened, remember a child class has access to the parent's functions and variables, and **child1** sets **self.x** because its **__init__**() function was automatically called; the parent's **__init__**() function was not called, and so **self.a** is undefined.

If we wish to use parent variables, their variables could be fixed, such as having b=8 (line 2, Figure 1.1). But what if we wanted to use a function to create a customizable plot layout that different child classes could use. Perhaps different classes wanted different plot sizes. In this case, it would be useful if the common design was maintained as a function in the parent class, but we'd like the child to be able to specify the size.

This can be achieved by having child classes call their parent's **__init__**() function as demonstrated with the child2 class (Figure 1.1 line 17). The command 'super().__init__()' invoked the parent's **__init__**() function. To test this, let's create a child2 object and test its variables (see Figure 1.4).

Now we see **self.a** is defined for the child because the parent's initialization was explicitly done using the super() command. Note, however, that while the current value of c2.x is 7, the parent class used a value of 9 for **self.x** but then **self.x** was updated after the parent was initialized.

It is also worth noting in our examples that when we create a child object, there is no parent object created; a child object can have features and capabilities associated with the parent class design, but there is no separate parent object. We can create a parent object and did in Figure 1.2, but we don't need to. In our examples, objects **p**, **c1**, and **c2** are different independent entities and the only things they have in common are definitions and design.

```
 1. c2=child2(7)
 2. Hi - this is the parent!
 3.
 4. c2.a
 5. Out[11]: 1
 6.
 7. c2.b
 8. Out[12]: 8
 9.
10. c2.x
11. Out[13]: 7
12.
```

FIGURE 1.4   Creating and examining a child2 object.

There is much more that could be said about OOD but what we have covered here is really all we need for the projects covered in this book. Our minimal usage is more than adequate for creating coding structures that are manageable and easily changed.

## A FEW PYTHON TIPS

Python is probably the most popular language used by data scientists and is obviously very powerful. This kind of power comes at a cost – complexity, and a learning curve. For scientists who don't necessarily have an extensive programming or computer science coursework background, there are some concepts and topics that are worth reviewing when dealing with the usual data structures, such as time series data and arrays, and a simplified summary of ones regularly encountered in our code examples will now be presented.

### Lists vs Arrays

Python has two kinds of structures for holding vectorial information: lists and NUMPY arrays. In Figure 1.5, we see how lists can be created and merged. Lists can also be indexed, so in this example, l3[2] is 3 – remember Python indexes start from 0.

But what if we wish to do something with a list? Suppose we want to operate on every list element? Perhaps multiply them by 3? This can be done using a very powerful list operation form shown in Figure 1.6.

But there is another way to achieve this by using Numpy arrays. For example, if we import the Numpy class as np using a command like '**import numpy as np**,' we could create a numpy array and simply multiply it by a multiplier. We can also combine numpy arrays mathematically. Examples of these capabilities are shown in Figure 1.7, where from the console, we create two lists, use them to create numpy arrays **x[]** and **y[]**,

```
 1. l1=[1,2,3]
 2.
 3. l2=[4,5,6]
 4.
 5. l3=l1+l2
 6.
 7. l3
 8. Out[17]: [1, 2, 3, 4, 5, 6]
 9.
10. l3.append(9)
11.
12. l3
13. Out[19]: [1, 2, 3, 4, 5, 6, 9]
14.
```

FIGURE 1.5  Working with lists.

```
1. l4 = [3*l for l in l3]
2.
3. l4
4. Out[21]: [3, 6, 9, 12, 15, 18, 27]
5.
```

FIGURE 1.6    Forming a new list l4 by looping through list l3 and multiplying each element by 3.

```
1. import numpy as np
2.
3. x = np.array(l1)
4.
5. y=np.array(l2)
6.
7. z = x + y
8.
9. z
10. Out[26]: array([5, 7, 9])
11.
12. list(z)
13. Out[27]: [5, 7, 9]
14.
```

FIGURE 1.7    Numpy arrays, unlike lists, allow for adding vectors in a traditional sense.

and then add the elements using **z = x+y**. We can recover the list form using the list() function.

It is worth noting that if we wish to rescale array elements, we can do it using the loop style shown in Figure 1.6 (line 1) or more simply by multiplying the array by a multiplier (see Figure 1.8 line 6)

An important and useful Numpy function is to create an empty array. For example, **np.zeros(6)** produces the array: [0,0,0,0,0,0].

What happens when we multiply a list by an integer? An example is shown in Figure 1.9.

```
1. zz = [3*i for i in z]
2.
3. zz
4. Out[29]: [15, 21, 27]
5.
6. zz = 3*z
7.
8. zz
9. Out[31]: array([15, 21, 27])
10.
```

FIGURE 1.8    Two ways to rescale array elements.

```
1. w = [1,0,0]
2.
3. 3*w
4. Out[35]: [1, 0, 0, 1, 0, 0, 1, 0, 0]
5.
```

FIGURE 1.9    Multiplying a list by an integer.

```
 1. z
 2. Out[36]: array([5, 7, 9])
 3.
 4. z[-2:]
 5. Out[37]: array([7, 9])
 6.
 7. z[2:]
 8. Out[38]: array([9])
 9.
10. z[1:2]
11. Out[39]: array([7])
12.
```

FIGURE 1.10    Slicing arrays.

The result is a repeated list – a very useful tool for creating lists based on repeated patterns such as weekdays, to annual ones, for modelling sampling distributions. Sometimes we only want to use part of an array. This can be done using the ':' operator as shown in Figure 1.10. On line 4, we take the last two elements using '[-2:],' all elements including and after element 2 '[2:],' and elements between indexes 1 and 2 '[1:2].' Note, that in Python indexing, a range such as '3:9' does not include the 9th element (see lines 10–11).

It can be very useful to be able to move back and forth between the worlds of lists and arrays. Many times we can simply use lists, but if working with data that is vectorial or matrix like, Numpy arrays are very convenient – especially if Numpy mathematical functions are needed.

## DataFrames

Pandas dataframes are two-dimensional matrices that can hold many different kinds of data types. They allow data scientists organize and manipulate multi-variate data. Newcomers to Python can be intimidated by them, often because they are a 'level above' simple arrays and can have non-numeric content. Some uses can rely on cryptic and dense syntax, but for our purposes, a straightforward usage will be sufficient. In this section, we will provide a summary of their main features and to serve as an initial overview for the novice, and as a reminder for the researcher who only deals with coding occasionally and can benefit from a short overview; and in doing so, hopefully establish the baseline understanding necessary for those wishing to adapt our later examples.

In Figure 1.11, a dataframe (**df**) is created one column at a time. Column 'x' is created from a list on line 5, and column 'y' from a list on line 6. Once created, a dataframe column can be accessed using either **df['x']**, or, since the column name is a simple name, by **df.x**

However, **df.x** is not a list as such and can be converted into a list using **df.x.tolist()**. On line 9, a new column is added by adding the two original

```
 1. import pandas as pd
 2.
 3. df = pd.DataFrame()
 4.
 5. df['x'] = [1,2,3]
 6.
 7. df['y'] = [4,5,6]
 8.
 9. df['z'] = df['x'] + df['y']
10.
```

FIGURE 1.11   A short code to create a dataframe using two lists.

```
 1. print(df)
 2.    x  y  z
 3. 0  1  4  5
 4. 1  2  5  7
 5. 2  3  6  9
 6.
```

FIGURE 1.12   Using the print() command to display the dataframe.

columns together. We could have implemented this in a simpler fashion: **df['z'] = df.x + df.y**. The resulting dataframe can be seen by typing **df** at the console or using print(df) (see Figure 1.12).

The first column without a header in Figure 1.12 is called the index and can be listed using **df.index.tolist()**, which would return a list **[0,1,2]**.

We have seen how to extract a column; how do we extract a row from a dataframe? We can do this using **iloc**, as shown in Figure 1.13, where **df.iloc[1]** extracts the row, and again, we can use **tolist()** to convert it into a list.

The **iloc** function is very powerful, and its argument can use up to two, comma separated lists used to find the dataframe elements that meet the row and column selectors:

**iloc[2]** would return row 2

**iloc[-1]** would return the last row

**iloc[:,2]** would return column 2

**iloc[:,-1]** would return the last column

**iloc[[1,2],[0,1]]** selects rows 1 and 2, and columns 0 and 1.

```
 1. df.iloc[1]
 2. Out[308]:
 3. x    2
 4. y    5
 5. z    7
 6. Name: 1, dtype: int64
 7.
 8. df.iloc[1].tolist()
 9. Out[309]: [2, 5, 7]
10.
```

FIGURE 1.13   Extracting row with index 1 using **df.iloc[1]**.

Remember, when using the ':' selector, such as [2:8], only the rows/columns from 2 through 7 are returned.

Sometimes we wish to extract values conditionally, for example, to find rows in a dataframe based on column values. For example, if we wished to extract rows in our dataframe here, where the y value was 5, we could do this with **df[df.y == 5]**.

To delete a column, such as column 'z,' do: **del df['z']** (interestingly, **del df.z** is not recommended).

We have barely scratched the surface here in reviewing dataframe basics, but this is most of what we need for later applications.

## CHAPTER SUMMARY

With our brief demonstration code, we have shown how subclasses can use resources from the parent, or customize them locally, and also how we can invoke subclasses and pass them parameters using the command line. We will use this kind of technique repeatedly throughout this book. It's a very primitive use of OOD but very effective for our purposes. A small amount of OOD goes a long way. In the next chapter, we will follow this strategy to allow us to build collections of models (subclasses) that can share functions from the parent resource class. In doing so, we will have a solution, where additional models could be easily included.

We also saw how we can have a choice between using lists and arrays, and we often need to be able to change from one to the other, but lists and arrays are very different. For vectorial problems, arrays should be used since they allow scaling and offsets to be applied to the arrays. With dataframe, we can be very functional and effective if we know the basic rules we described here, in how to create dataframes from lists, how to convert columns to lists, and how to extract data based on either integer locations (**iloc**) or conditional matching. These tools will be relied on in later chapters. In the next chapter, we will explore the first set of models and simulations, where we will investigate gravity fields for particle distributions we assemble.

## CHAPTER 2

# Exploring Data

S CIENTISTS WORK WITH DATA since without data, theories could not be tested and revised. As we develop our understanding of science, knowledge of theory must be matched with an appreciation of how to analyze data and how to create models to test with the data. While this is the general theme throughout this book, in this chapter, we will show how powerful even simple polynomial equations can be for modelling fundamental theories and will show how graphical presentations can lead to striking conclusions and how they will serve as a good introduction to Python's plotting capabilities.

Consider a very simple observation: The Earth orbits the Sun in about 365 days. We now know there is an equilibrium at play, between gravitational forces, the masses of the Sun and Earth, the Earth's velocity (speed and direction), and the distance between them. But there are other possible factors: their sizes and rotation speeds; their composition and surface temperatures. Perhaps you think I'm over-stating things a little. How could temperature and composition be factors? Well consider this, because of the Sun's high surface temperature and luminosity, it creates pressure from photons and emits boiled off particles. As a result, a mylar balloon placed at the Earth's distance from the Sun will not orbit in 365 days; it will be pushed outward and might escape the Solar System because of radiation pressure. So, gravity is not always the dominant force.

The point I'm making here is that real-world problems might have many factors to consider, and the scientist or engineer needs to be able to decide which are important and which are not, so the simplest possible descriptions (laws) can be developed that explain the observations. Ultimately,

 DOI: 10.1201/9781003600046-2

they are trying to identify patterns, relationships, between variables, and there is absolutely no excuse for undertaking such a search without exploring how various observations/variables interact, by examining plots of one against another. If relationships are found (i.e., the plots are not just of randomly scattered points), then a relationship is being shown between the variables; better yet, would be to be able to say whether the relationship was universal and relevant to all systems, and not the one being investigated.

## KEPLER'S THIRD LAW

When Kepler considered the problem of planetary motion, he drew the conclusion that planetary orbits obeyed a rule relating their periods and their distances from the Sun. He didn't know the sizes or temperatures (except for the Earth), and the ideas of mass and gravity had to wait for Isaac Newton, so he wasn't confused or distracted from fundamental properties: Period ($\mathbf{P}$) and distance from the Sun – the orbit semi-major axis ($\mathbf{a}$). Kepler saw there was a pattern and stated that the orbit period and orbit size obeyed the rule (we now call Kepler's third law)

$$\mathbf{P}^2 = k\,\mathbf{a}^3 \tag{2.1}$$

where k is a constant of proportionality and equals 1 if time is measured in Earth years and distance is measured in Earth-Sun units (called the Astronomical Unit or AU). We could rewrite the equation as:

$$\mathbf{P} = \mathbf{a}^{3/2} \tag{2.2}$$

It is easier to understand his reasoning by looking at the data graphically. In Figure 2.1 the top panel shows $\mathbf{P}$ vs $\mathbf{a}$ for the planets Mercury through Saturn (Kepler wouldn't have known about the others). The middle panel shows a linear and quadratic curve drawn to the same x and y axis limits which shows the planets operate between these two extremes. The third panel shows the planets with the linear and quadratic curves, and a curve with exponent 1.5 – which is seen to exactly match the planetary data, so Kepler's third law works!

If we didn't know about Kepler's third law, we could have used this approach to discover it, by trying curves with different exponents until we found the best one through trial and error. This is essentially what Kepler did – he had no underlying theory of physics to justify his results.

Verifying Kepler's 3rd Law

Solar System Planets (raw data): P vs a

Comparisson Power Law Curves

Raw data and power-laws 1, 1.5, and 2.0

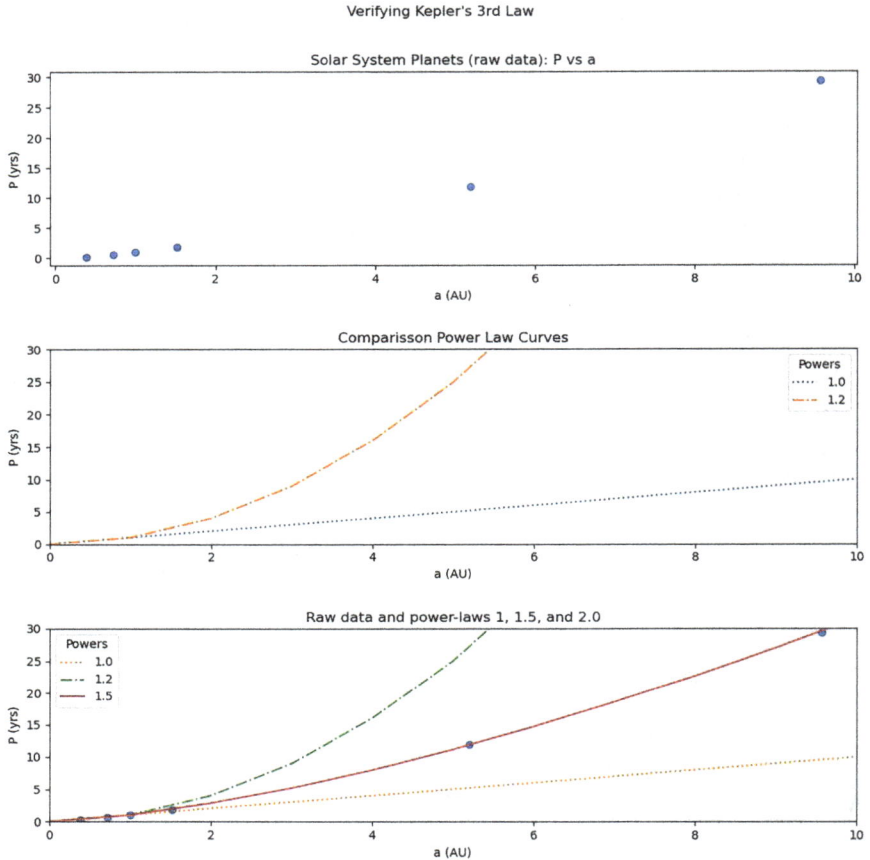

FIGURE 2.1    Verifying Kepler's third law. **P** vs **a** values clearly follow some well-defined increasing trend (top). For comparison (middle), linear and quadratic curves are drawn to the same scale and clearly bound the planetary data. Combining the plots (bottom), we find a curve with exponent 1.5 (as stated in Kepler's third law) which exactly matches the planetary data.

It is worth noting that 'coincidences' should always get our attention. Kepler found the period depended on **a** to the 1.5 power, not 1.49, or 1.47, or 1.53. Is it a coincidence that the needed power was exactly 1.5 or equivalently the ratio of two integers: 3/2? The answer is it is not a coincidence, and while Kepler might have used 1.5 on the assumption it was exactly right, strictly speaking this was supposition on his part. It was only when Newton used his famous law of gravity when studying planetary dynamics which resulted in a derivation of Kepler's laws, was it known that yes indeed, **P** depends on **a** to the power of 1.5. Why would the law of gravity cause this? The answer is because Newton's law of gravity itself builds in an integer power of 2 by saying the gravitational force depends on $1/r^2$ which could be re-written as $4\pi/4\pi r^2$, and since the denominator is just the area of a sphere that power of 2 is not an approximation; it is the area of a sphere, which is a perfect power law, hence gravity depends exactly on the power of 2.

When analyzing data, it is important to appreciate the role played by the scales of things. In Figure 2.1, we used units where periods were in years and distances in AUs. That was human bias. If we grew up on Jupiter, a scientist there might want the period for Jupiter to be one. Would that make any difference? This is easy to test, let's rescale the X and Y axes by constants **B** and **A**, respectively, so Equation 2.3 becomes:

$$AP = (Ba)^{1.5} \tag{2.3}$$

and hence,

$$P = \frac{B^{1.5}}{A} a^{1.5} \tag{2.4}$$

So,

$$\mathbf{P} = \mathbf{K}\,\mathbf{a}^{1.5} \tag{2.5}$$

In other words, in rescaling, all we effectively did was to introduce a different constant multiplier, **K**. What if we only rescaled one axis? It wouldn't change our conclusion since then either **A** or **B** would be 1.

But what about other systems, such as the moons of Jupiter? Do they obey Kepler's third law? Yes, and we could show it using the same methods we just used. However, before doing so, we must point out that unknown to Kepler, there was another important physical parameter present, namely the sum of the central and orbiting masses (M+m), as shown in Equation 2.6.

$$P^2 = (M+m)a^3 \tag{2.6}$$

For the Solar System, because the Sun is so massive, the sum of a planet's mass and the Sun's is essentially the same as the Sun's, and in our system of units using the year and the AU, the Sun's mass is 1. This hidden mass dependence needs to be taken into account, and for this reason, we cannot simply combine systems with different central masses, because each system has its own constant of proportionality, which depends on the central mass (mainly) and the unit choices. This means that we can compare different systems and show they all obey Kepler's third law, if we divide the (**P**, **a**) values for each system by any value pair from that system, since by doing this, we divide out the constants of proportionality and the mass terms in each system.

We will now demonstrate this by considering the four Galilean Moons, named after Galileo who after observing them move, famously concluded that not everything orbited the Earth – a devastating criticism of the widely believed Geocentric Universe cosmology. Again, we can overlay the data for the Galilean moons and the planets when each set is normalized to one of its members to adjust for each system having a very different central mass (see Figure 2.2), and because of this both sets will contain

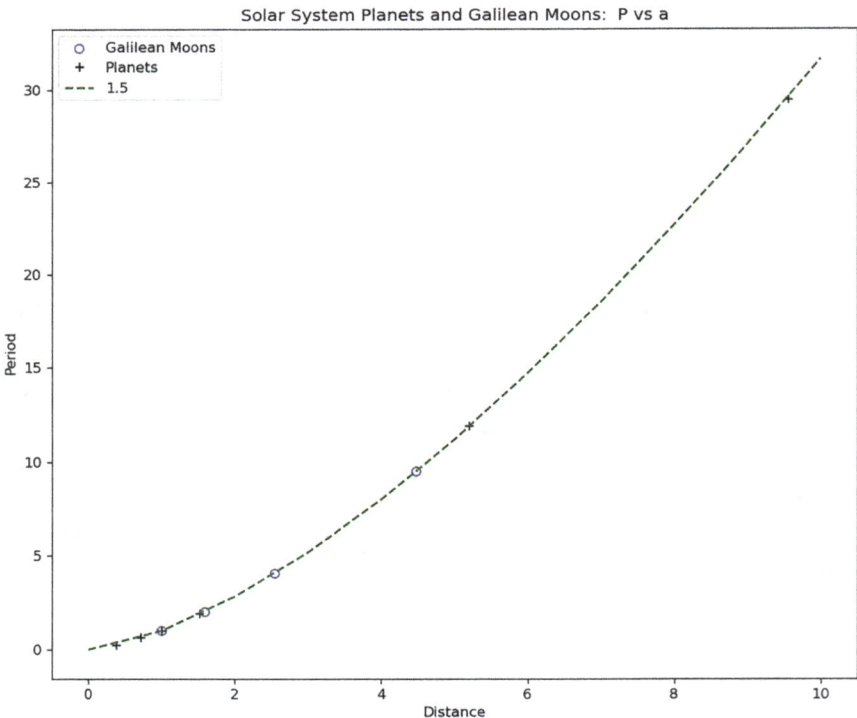

FIGURE 2.2 **P** vs **a** for the planets ('+' markers), along with the 4 Galilean moons 'o' markers).

a common point (1,1). Since both systems show obedience to the 1.5 power-law, Kepler's third law applies to both, as does Newton's law of gravity.

Because objects in the planetary (Solar System) and Jupiter orbiting systems are each scaled by one member of each, their constants of proportionality are set to one, and they all obey the same power-law rule.

There is another elegant way to view our data, namely by plotting log(**P**) vs log(**a**) since on a log-log plot, data points obeying a power line fall on a straight line, where the line's slope equals the exponent. This is shown in Figure 2.3 where the planets fall on one line, and the Galilean moons on another, but their respective lines are parallel and have a slope of 1.5, again showing planets and Jupiter's satellites obey Kepler's third law. Using this technique, we don't need to normalize the systems to take the mass effect into account, all systems obeying the 1.5 power law will be parallel, demonstrating the law's universality.

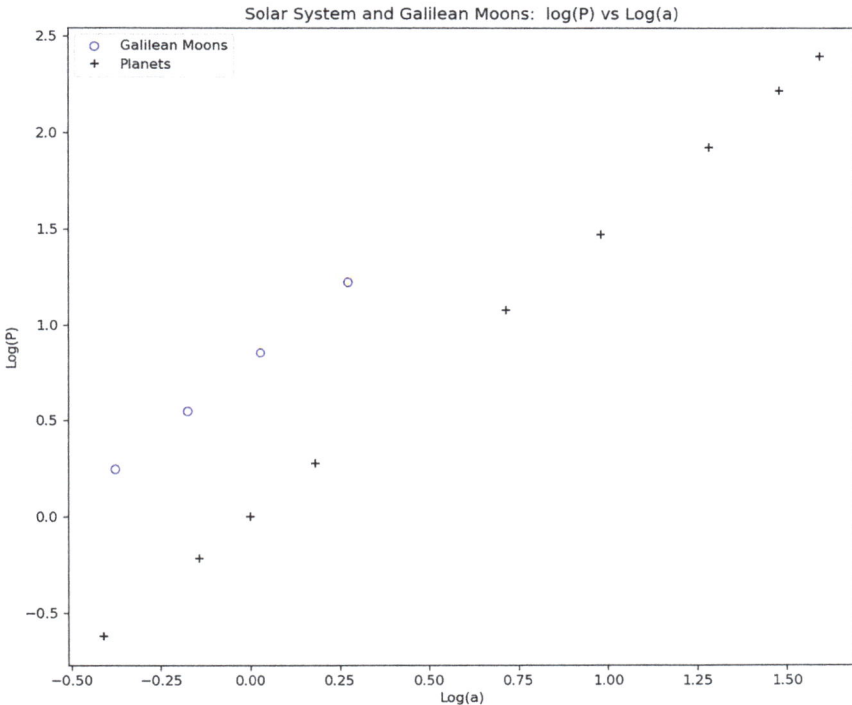

FIGURE 2.3   Power law relationships become straight lines when plotted on a log-log plot. Here, for the Galilean moons, and the planets, their log(**p**) vs log(**a**) values are plotted. Because each system obeys Kepler's third law, both systems fall on straight lines of slope 1.5.

## CLASS K3L PROGRAMMING NOTES

In demonstrating Kepler's third law here, we took advantage of Python's Matplotlib library and created a Python class (called **K3L**, in a file called **K3L.py** and shown in Figure 2.4) to hold the functions we used to create the figures here. Note some of the specific plotting capabilities used:

- Creating a panel of plots using the **subplot()** function (line 84)

```
1. import math
2. import matplotlib.pyplot as plt
3.
4.
5. class K3L:
6.     def __init__(self):
7.         self.make_objects()
8.         self.xmax = 10
9.         self.ymax = 30
10.
11.     def make_objects(self):
12.         self.nlist = ['Mercury','Venus','Earth','Mars','Jupiter', \
13.                       'Saturn', 'Uranus', 'Neptune', 'Pluto']
14.         self.alist = [0.39, 0.72, 1, 1.52, 5.2, 9.57, 19.2, 30.2, 39.2]  # AU
15.         self.plist = [.24, .61, 1, 1.88, 11.9, 29.4, 83.7, 163.7, 247.9] #Yrs
16.
17.         self.jnlist = ['Io','Europa','Ganymede','Callisto']
18.         self.jalist = [0.42, 0.67, 1.07, 1.88]   # million km
19.         self.jplist = [1.76, 3.52, 7.15, 16.7]   # days
20.
21.     def plot_raw_data(self):                             # subplot 1
22.         plt.plot(self.alist[0:6], self.plist[0:6],'o')
23.
24.         plt.show()
25.
26.     def plot_polys(self):                                # subplot 2
27.         xlist = list(range(0,11))
28.         p1 = [x    for x in xlist]
29.         p2 = [x**2 for x in xlist]
30.         plt.plot(xlist, p1,linestyle='dotted', label='1.0')
31.         plt.plot(xlist, p2,linestyle='dashdot', label='1.2')
32.         self.set_plt_lims(plt,0,self.xmax,0,self.ymax)
33.
34.     def plot_raw_plus_poly(self):                        # subplot 3
35.         plt.plot(self.alist[0:6], self.plist[0:6],'o')
36.         self.plot_polys()
37.         xlist = list(range(0,11))
38.         p15  = [x**1.5 for x in xlist]
39.         plt.plot(xlist, p15,linestyle='solid', label="1.5")   # add 1.5 power-law
40.         plt.show()
41.
42.     def planet_and_galilean_moons_scaled_plot(self):
43.         mydpi=120
44.         fig = plt.figure(figsize=(1200/mydpi,1000/mydpi),dpi=mydpi)
45.         plt.title("Solar System Planets and Galilean Moons:  P vs a")
46.         plt.xlabel('Distance')
47.         plt.ylabel('Period')
48.         pj = [p/1.76 for p in self.jplist]       # rescale by dividing by a point
49.         aj = [a/0.42  for a in self.jalist]
50.         plt.scatter(aj[0:7], pj[0:7],facecolors='none',edgecolor='blue', \
51.                     label='Galilean Moons')
52.         xlist = list(range(0,11))
53.         plt.plot(self.alist[0:6], self.plist[0:6],'+',color='black',label='Planets')
54.         xlist = list(range(0,11))
55.         p15  = [x**1.5 for x in xlist]
56.         plt.plot(xlist, p15,linestyle='dashed', color='green', label="1.5")
57.         plt.legend()
```

FIGURE 2.4    Class K3L.                                      *(Continued)*

```
58.          plt.show()
59.          plt.savefig('./Fig 2.2.jpg',dpi = mydpi)
60.
61.     def log_log_planet_and_galilean_moons_plot(self):
62.          mydpi=120
63.          fig = plt.figure(figsize=(1200/mydpi,1000/mydpi),dpi=mydpi)
64.          plt.title("Solar System and Galilean Moons:  log(P) vs Log(a)")
65.          plt.xlabel('Log(a)')
66.          plt.ylabel('Log(P)')
67.          pj = [math.log10(p)  for p in self.jplist]
68.          aj = [math.log10(a)  for a in self.jalist]
69.          plt.scatter(aj[0:7], pj[0:7],facecolors='none',edgecolor='blue',\
70.                        label='Galilean Moons')
71.          ap = [math.log10(x0) for x0 in self.alist]
72.          pp = [math.log10(p0) for p0 in self.plist]
73.          plt.plot(ap, pp,'+',color='black',label='Planets')
74.          plt.legend()
75.          plt.show()
76.          plt.savefig('./Fig 2.3.jpg',dpi = mydpi)
77.
78.
79.     def set_plt_lims(self,plt,xmin,xmax,ymin,ymax):
80.          ax = plt.gca()
81.          ax.set_xlim([xmin, xmax])
82.          ax.set_ylim([ymin, ymax])
83.
84.     def make_planet_panel(self):
85.          mydpi=100
86.          fig = plt.figure(figsize=(1200/mydpi,1200/mydpi),dpi=mydpi)
87.          fig.subplots_adjust(wspace=.1, hspace=.5)
88.          fig.suptitle("Verifying Kepler's 3rd Law")
89.
90.
91.          plt.subplot(3, 1, 1)
92.          plt.xlabel('a (AU)')
93.          plt.ylabel('P (yrs)')
94.          plt.title("Solar System Planets (raw data): P vs a")
95.          self.plot_raw_data()
96.
97.          plt.subplot(3, 1, 2)
98.          plt.xlabel('a (AU)')
99.          plt.ylabel('P (yrs)')
100.         plt.title("Comparisson Power Law Curves")
101.         self.plot_polys()
102.         plt.legend(title='Powers')
103.
104.         plt.subplot(3, 1, 3)
105.         plt.xlabel('a (AU)')
106.         plt.ylabel('P (yrs)')
107.         plt.title("Raw data and power-laws 1, 1.5, and 2.0")
108.         self.plot_raw_plus_poly()
109.         plt.legend(title='Powers')
110.
111.         plt.show()
112.         plt.savefig('./Fig 2.1.jpg',dpi = mydpi)
113.
114. if __name__ == '__main__':
115.     k3l = K3L()
116.
117.     k3l.make_planet_panel()
118.     k3l.planet_and_galilean_moons_scaled_plot()
119.     k3l.log_log_planet_and_galilean_moons_plot()
120.
```

FIGURE 2.4 (CONTINUED)    Class K3L.

- Specifying saved plot sizes/resolutions and names (e.g., lines 43–44, and 59)

- Choosing different kinds of lines (**dashed** and **dotted**) (e.g., lines 30–31)

- Using labels so legends could be added (e.g., lines 70 and 73)

- Adding titles to plots (e.g., line 45)

- Setting axis limits (lines 81–82)

- In some instances, the **plt.scatter**() function gives more flexibility than **plt.plot**() and was used to create circle markers instead of dots (line 50)

- When creating panels of subplots, the spacings might be fussy and require width and height spacings be specified. Generally, the function **tight_layout**() will achieve acceptable results and should be used before trying to solve the layout problem manually, however, using the **wspace** and **hspace** settings can be very effective when **tight_layout**() doesn't work (see line 87).

Pay particular attention to Matplotlib's line-style capability, since documents and graphics you produce might end up in black and white, with no color, and being able to differentiate between curves and plots, based on something other than color could be important.

When modelling, it is often very useful to be able to create a list of values, such as a list ranging from 0 to 10: **xlist = list(range(0,11))**. Remember in Python, the **range(0,11)** function will return the sequence '0,1,...10,' and the **list**() function is needed to turn it into a Python list. Another instance of Python indexing was used to select a subset of the planets limiting us to those Kepler would have used, namely the first six planets. This was done when sub listing the **alist** and **plist** lists using **alist[0:6]** and **plist[0:6]** at line 22 for example.

It is also very useful to be able to derive one list from another. For example, given **xlist[]**, then the list **y[]** of x-squared values could be found from: **y = [x**2 for x in xlist]** (see lines 28–29).

Note also, we used a very useful and clever trick at the end of the class definition file (line 114). We added an if-condition (**if __name__ == '__main__':**) which will be true, if the **K3L.py** is itself run as a standalone code and any commands following this if-condition will be run. Otherwise, if the **K3L** class is invoked by another program, the demonstration code following the if-condition is ignored. Using this strategy allows you to develop and share classes with embedded demonstration code that can be activated if needed. These kinds of techniques will be used repeatedly throughout this book.

## THE YALE BRIGHT STAR CATALOG AND THE HR DIAGRAM

Astronomers love to make catalogs. With a data catalog, they have a collection of object data that supports statistical analysis and data manipulation. Some catalogs can be enormous and difficult to work with because they can also have too much information, and so it is useful for students to know how to find and extract the data they really need. Here we will show how to access and download subsets of data from a NASA site, and conduct some investigations of the data. Our primary goal is to create an Hertzsprung-Russel (HR) diagram from the data, which is a plot of luminosity vs temperature (or spectral type). We will estimate the luminosity from the magnitude and parallax, and the temperature will be based on the spectral type.

We will get our data from heasarc.gsfc.nasa.gov and will use its filtering capability to extract parts of the Yale Bright Star Catalog. We want key properties such a star's visual magnitude to tell us about its brightness; the spectral type, about its temperature and class (i.e., whether it is a main sequence star or not); and the parallax about its distance. Knowing the parallax and brightness allows us to know the intrinsic energy output (luminosity or absolute magnitude). This catalog, even though it only contains relatively bright stars, is also useful for providing data if a user wanted to do other projects such as creating constellation charts, perhaps as a background for asteroid and planet finder charts.

The https://heasarc.gsfc.nasa.gov/db-perl/W3Browse/w3table.pl?Mission Help=star_catalog page offers a list of catalogs, and we will use its Bright Star Catalog (**BSC5P**), 5th edition, because it has a manageable number of stars (about 9000) and has the data columns we need. On the web site page's top left corner is a 'Browse' link to the page used to extract data, and this offers a comprehensive set of choices. Since we only need a subset, we show the fields we selected in Figure 2.5 (identifiers, position, and magnitude), and we also show the part specifying the output format. Since this is a small dataset, only about 9000 lines, we decided to not limit the output rows and we selected Excel format, since we would later like to use Excel to create a **csv** version of the data. On clicking the Start Search button, the data was downloaded to my PC to a file called 'browse_results.xls' in my

Main Search Form                    Search of STAR CATALOG Catalog(s)

| Description | Catalog Data | Default Radius (arcmin) | Mission | Table Type |
|---|---|---|---|---|
| Bright Star Catalog | bsc5p N | 1 | STAR CATALOG | Object |

**1. Enter any constraints on the query below. [ Help on constraint syntax ]**
Examples of query constraints:
(What about wildcards, spaces, and case sensitivity?)

**2. To change the fields that are returned, select the box in the "View" column beside each field desired.**

**3. To sort the results by any field, select one box in the "Sort" column beside the field to sort on.**

| View □ All | Sort | Parameter (Unit) | Query Terms | Min Value | Max Value | Value Type |
|---|---|---|---|---|---|---|
| ☑ | ○ | name | | HR 1 | HR 999 | string |
| ☑ | ○ | alt_name | | 1 Aqr | Zet2Sco | string |
| ☑ | ○ | ra | | 00 00 19.20 | 23 59 55.01 | position |
| ☑ | ○ | dec | | -88 57 23.0 | +89 15 51.1 | position |
| ☑ | ○ | vmag | | -1.46 | 7.96 | float |
| ☑ | ○ | multiple | | A | W | string |
| ☑ | ○ | class | CV CLASSICAL NOVA / GLOBULAR CLUSTER EXTENDED GALACTIC OR EXTRAGALACTIC / OPEN STAR CLUSTER | | | |
| ☑ | ○ | paralax (arcsec) | | -0.032 | 0.751 | float |
| □ | ○ | pmdec (arcsec/yr) | | -5.813 | 3.208 | float |
| □ ads | ○ | pmra (arcsec/yr) | | -3.646 | 4.136 | float |
| □ ads_comp | ○ | ra_1900 (degree) | | 0.00000 | 359.98417 | float |
| □ bii (degree) | ○ | ra_1950 (degree) | | 0.00610 | 359.98280 | float |
| | ○ | radvel (km/s) | | -140 | 289 | integer |
| | ○ | radvel_comm | | ? | V? | string |
| | ○ | ri_code | | | E | string |
| | ○ | ri_color | | -0.53 | 3.71 | float |
| | ○ | rotvel (km/s) | | 0 | 455 | integer |
| | ○ | rotvel_comm | | < | =< | string |
| | ○ | rotvel_uncert | | | v | string |
| | ○ | sao | | 106 | 258996 | integer |
| | ○ | spect_code | | e | | string |
| ☑ | ○ | spect_type | | A/FmDel Del sgK2 | | string |
| □ | ○ | ub_color | | -1.11 | 7.40 | float |

**Limit Results To:** [No Limit ▾] rows
**Output Format:** [Excel-compatible ▾]
**Show All Parameters:** □ Select to display all catalog parameters instead of only defaults.

FIGURE 2.5   Close ups of the selection options used for our download where the full catalog is used and the download format chosen.

Downloads folder, and Figure 2.6 shows what the first 18 lines of the 9111 look like when opened in **Excel**. This is a useful dataset for physics or astronomy students to have. Note the downloaded file has two workbook pages. The first can be deleted.

To build an HR diagram, we only need three of the columns, and we will reject rows flagged as being multiple stars and rows without parallax measurements. To do this, first select the whole table in Excel, sort on the

| | A | B | C | D | E | F | G | H | I | J | K |
|---|---|---|---|---|---|---|---|---|---|---|---|
| 1 | name | alt_name | ra | dec | vmag | multiple | class | bv_color | parallax | spect_cod | spect_type |
| 2 | HR 7228 | Sig Oct | 317.1925 | -88.9564 | 5.47 | | STAR F0III | 0.27 | | | F0III |
| 3 | HR 8294 | | 341.3692 | -88.8183 | 6.57 | | STAR F0IV | 0.28 | | | F0IV-V |
| 4 | HR 5491 | | 232.0796 | -88.1331 | 6.48 | | STAR A0 | 0.3 | | | Am |
| 5 | HR 6721 | Chi Oct | 283.6954 | -87.6058 | 5.28 | | STAR K3III | 1.28 | | | K3III |
| 6 | HR 6133 | | 258.9971 | -87.5664 | 6.57 | | STAR G5III | 0.91 | | | G5III |
| 7 | HR 8862 | Tau Oct | 352.0154 | -87.4822 | 5.42 | | STAR K2III | 1.27 | | | K2III |
| 8 | HR 2848 | | 101.7446 | -87.025 | 6.47 | | STAR F3V | 0.42 | | | F3V |
| 9 | HR 6139 | | 255.2437 | -86.3644 | 6.04 | | STAR A2V | 0.05 | | | A2V |
| 10 | HR 4709 | | 186.4062 | -86.1506 | 6.33 | | STAR K0III | 1.08 | | | K0III |
| 11 | HR 8505 | Ups Oct | 337.9058 | -85.9672 | 5.77 | | STAR K0III | 1.02 | | | K0III |
| 12 | HR 5084 | Kap Oct | 205.2312 | -85.7861 | 5.58 | | STAR A0 | 0.18 | | | Am |
| 13 | HR 3678 | Zet Oct | 134.1712 | -85.6631 | 5.42 | | STAR A8IV | 0.31 | | | A8-9IV |
| 14 | HR 4595 | | 180.5838 | -85.6317 | 6.05 | W | STAR K3III | 1.29 | | | K3III |
| 15 | HR 1271 | | 55.6338 | -85.2622 | 6.41 | W | STAR B9IV | -0.01 | 0.008 | | B9.5IV |
| 16 | HR 6552 | | 270.3921 | -85.2147 | 6.45 | | STAR F5IV | 0.44 | | | F5IV |
| 17 | HR 4870 | Iot Oct | 193.7442 | -85.1233 | 5.46 | W | STAR K0III | 1.02 | | | K0III |
| 18 | HR 47 | | 3.3308 | -84.9942 | 5.77 | | STAR M0III | 1.72 | | | M0-1III |

FIGURE 2.6   The downloaded file contains 9111 rows – we are showing the first 18 here.

**multiple** column (column F), and delete all rows with a non-blank 'multiple' field entry. There should be 1578 of these. After this, the 'multiple' column can be deleted. Repeat this process to sort by parallax and delete rows without a parallax entry, leaving us with a sample size of 2464 stars to work with.

Now that we have our data, we use **Excel** to save it as a csv file called **BSC5.csv**.

To import our data into our Python code, we can work with Pandas dataframes. A dataframe is a matrix that can hold many different kinds of data, and the Pandas library has many different features to allow for dataframe manipulation. For most of the projects we tackle in this book, we don't need the power of dataframes, but in this case, they are well worth using. For example, to manage the BSC data, we started by creating a BSC class definition and one of the first things we had it do was import the catalog, as shown in Figure 2.7.

Once run, an instance of the class is created (line 16) as **bsc**, and a preview of the **bsc.df** dataframe can be displayed on the console with a command like '**print bsc.df**' or, use the IDE's variable explorer to examine its contents (as shown in Figure 2.8).

We only need the **vmag**, **class**, and **parallax** columns since these can be used to estimate the absolute magnitude (**M**), the luminosity (**L**), and the temperature (**T**), since an HR diagram is a plot of **L** (or **M**) against **T** (or spectral type).

Knowing a star's apparent magnitude **vmag** (or **m**) and parallax **p**, the absolute magnitude is estimated from:

$$\mathbf{M} = \mathbf{m} - 5\log_{10}\left(\mathbf{d}/10\right) \tag{2.7}$$

```
 2. import matplotlib.pyplot as plt
 3.
 4.
 5. class BSC:
 6.     def __init__(self):
 7.
 8.         self.df = pd.DataFrame()
 9.         self.read_bsc()
10.
11.     def read_bsc(self):
12.         self.df = pd.read_csv('bsc5.csv')
13.
14.
15. if __name__ == '__main__':
16.     bsc = BSC()
17.
```

FIGURE 2.7 Importing the **bsc5.csv** file into a pandas dataframe take very little code when using the pandas **read_csv** function.

| Index | name | alt_name | ra | dec | vmag | class | bv_color | parallax | spect_code | spect_type |
|-------|------|----------|-----|-----|------|-------|----------|----------|------------|------------|
| 163 | HR 776 | Mu Hyi | 37.9188 | -79.1094 | 5.28 | STAR G8III | 0.98 | 0.001 | nan | G8III |
| 164 | HR 852 | Nu Hor | 42.2562 | -62.8067 | 5.26 | STAR A2V | 0.1 | 0.001 | nan | A2V |
| 165 | HR 7968 | Iot Ind | 312.875 | -51.6083 | 5.05 | STAR K1III | 1.13 | 0.001 | nan | K1III-III |
| 166 | HR 4476 | nan | 174.391 | -47.7472 | 5.44 | STAR K2III | 1.24 | 0.001 | nan | K2III |
| 167 | HR 2618 | 21Eps CMa | 104.656 | -28.9722 | 1.5 | STAR B2II | -0.21 | 0.001 | nan | B2II |
| 168 | HR 7980 | 18Ome Cap | 312.955 | -26.9192 | 4.11 | STAR M0III | 1.64 | 0.001 | nan | M0-III-IIIbBa0.2 |
| 169 | HR 7066 | nan | 281.871 | -5.705 | 5.2 | STAR K0I | 1.47 | 0.001 | v | K0Ibp |
| 170 | HR 1895 | 41The1Ori | 83.8187 | -5.3897 | 5.13 | STAR O6 | 0.02 | 0.001 | v | O6p |
| 171 | HR 1896 | 41The1Ori | 83.8221 | -5.3878 | 6.7 | STAR B0V | 0.09 | 0.001 | nan | B0.5Vp |
| 172 | HR 1893 | 41The1Ori | 83.8163 | -5.3872 | 6.73 | STAR B0V | 0.02 | 0.001 | nan | B0.5V |
| 173 | HR 1894 | 41The1Ori | 83.8171 | -5.3853 | 7.96 | STAR B0V | 0.24 | 0.001 | nan | B0V |
| 174 | HR 8251 | nan | 323.823 | -3.9831 | 5.77 | STAR | 1.11 | 0.001 | nan | gG9 |

FIGURE 2.8 Using the Python Spyder IDE's variable explorer, we can see our **bsc.df** dataframe's contents.

where **d**, the distance, measured in parsecs, is simply 1/**p**. Note, only 2400 entries in the catalog had parallax information and were useful for our purpose here.

Estimating temperature is a little trickier. Looking at the class column in our dataframe, we see string type entries consisting of an object type (STAR) and the spectral information. For example, the first entry shows the object is G8III, this means it is G8 spectral subclass object of luminosity class III. Luminosity classes tell us the kind of star we are dealing with, and the major kinds use roman numerals I–V. Ordinary stars in their adult, hydrogen core burning phase are of type V.

Spectral classes follow the famous 'Oh Be A Fine Girl/Guy Kiss Me' mnemonic to remind us of the order O-B-A-F-G-K-M from the hottest to the coldest stars. (The historical reason for this is spectra were first labelled alphabetically and then rearranged to better follow their patterns, and it was only later that the order was recognized as showing decreasing temperature, but the order had been established by then, and so the HR diagram always plots decreasing temperature or spectral type on its x-axis.)

Spectral types are subdivided 0–9, with 0 being the hottest, and so on, and each spectral type has temperature limits. We will use a linear approximation between these limits, so, for example, since the G spectral type spans a 700K range between 5300K and 6000K, a G0 star will be 5300K, and a G2 would be 5300K + 0.2* 700K = 5440K.

Class **BSC** was developed to show major features of the dataset and was used to generate Figures 2.9 and 2.10. For these figures, luminosity classes I–IV were colored red, and class V colored blue. In Figure 2.9 the apparent

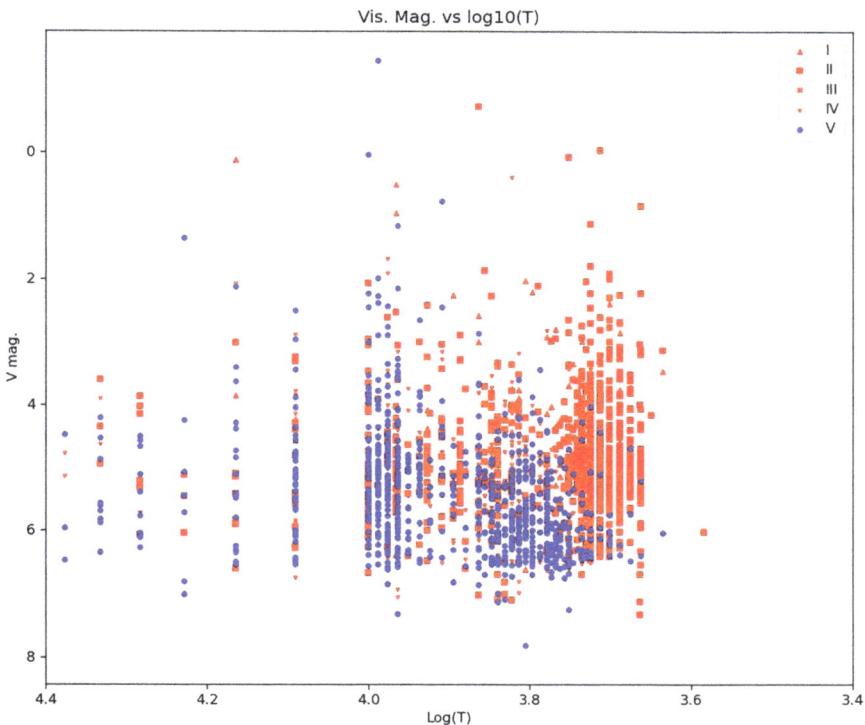

FIGURE 2.9   A plot of apparent magnitude against log(T) shows low temperature stars are mainly giants/super-giants, but the magnitude itself cannot differentiate between them.

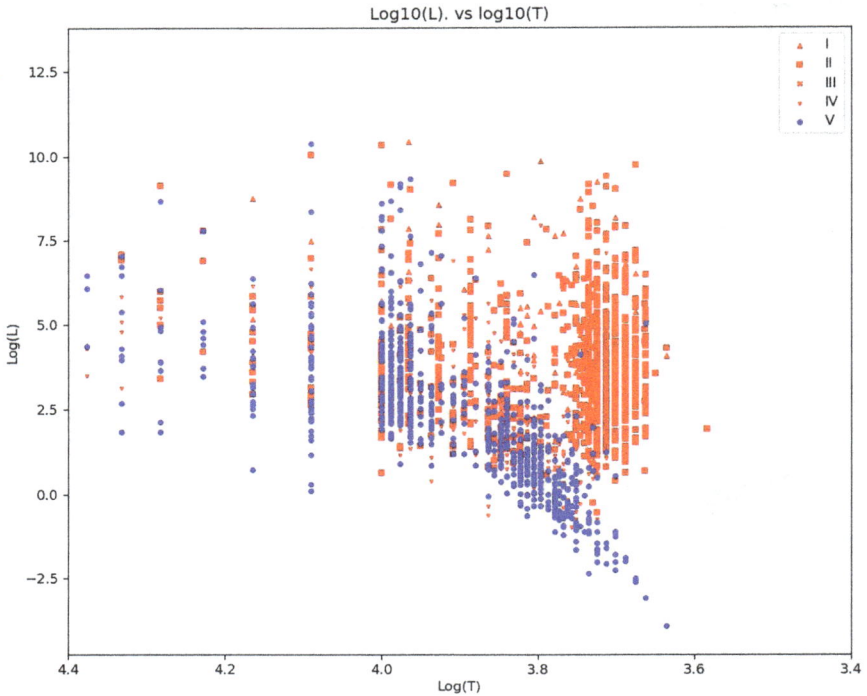

FIGURE 2.10  Plotting log(L) against log(T) shows the class V, Main Sequence stars have greater luminosities at higher temperatures. The slope changes from being almost flat at high temperatures to a maximum at low temperatures.

magnitude (**m**) was plotted against log(**T**), while in Figure 2.10, log(**L**) was plotted against log(**T**). (Note, since the magnitude scale is intrinsically logarithmic, it was not necessary to plot the log of the apparent magnitude in Figure 2.9.) Both figures show a separation between spectral class groups, and we refer to the class V objects as being on the 'Main Sequence,' while the others, especially classes I-III, are giant/super-giant stars. We now know Main Sequence stars are in their adult life (hydrogen core burning) stage.

In Figure 2.10, the Main Sequence stars display a slope which demonstrates that luminosity depends on temperature. From elementary astrophysics, we know the luminosity should depend on $R^2T^4$, where $R$ is the star radius, the slope of the Main Sequence in Figure 2.10's log-log plot should be 4, if all stars had a constant radius. A visual inspection shows it can be much greater than this, which is telling star sizes are increasing significantly with temperature.

So, in using apparent magnitude and temperature, we could detect different luminosity classes, but we needed the absolute magnitude or luminosity to show Main Sequence stars had greater luminosities at higher temperatures. In other words, apparent magnitude (brightness) could not distinguish between class or among Main Sequence stars, but absolute magnitude could differentiate among Main Sequence stars.

However, there is roughness or granularity in our plots because we inferred temperatures from grouped data, the spectral subclass categories, instead of from directly measured temperatures, and while we can see the prominent Main Sequence (class V stars) we will need to use better data if we wish to better explore the underlying astrophysics, and for that we will use the DEBCat data below.

## CLASS BSC PROGRAMMING NOTES

The code used to generate Figures 2.9 and 2.10 is shown in Figure 2.11.

The code relies on the PANDAS library and on its dataframe structures to support data input and management. Among the key dataframe features used were:

Reading the csv data file into a local dataframe (line 15)

Excluding/filtering dataframe rows (lines 16 and 91)

Renaming dataframe column names (line 17)

Extracting a dataframe column into a list (line 20)

Adding a list as a new column in a dataframe (line 23)

The class is initialized using the usual __**init**__(**self**) function and creates lists for variables to be used, and also sets the colors and markers for plotting the luminosity classes (lines 11–12).

The star catalog is imported by the **read_bsc** function, and at line 21, a distance column built from parallax using the formula **d=1/p**. Note the very useful and compact strategy for operating on a list where we calculate a list of distances from a list of parallax values: **d = [1/p0 for p0 in p]**.

New columns are created in the dataframe for absolute magnitude, luminosity, and estimated temperature by function **get_MLT_values()**.

Spectral subclasses are mapped to temperatures (line 26) by creating a dictionary by function **build_spectral_type_temperature_dictionary()**.

```
 1. import pandas as pd
 2. import matplotlib.pyplot as plt
 3. import math
 4.
 5.
 6. class BSC():
 7.     def __init__(self):
 8.
 9.         self.sp_temps = {}
10.         self.df       = pd.DataFrame()
11.         self.col_map  = {'I':'r', 'II':'r', 'III': 'r','IV':'r','V':'b'}
12.         self.mrk_map  = {'I':"^", 'II':",', 'III': "x",'IV':"1",'V':"o"}
13.
14.     def read_bsc(self):
15.         self.df       = pd.read_csv('bsc5.csv')
16.         self.df       = self.df[self.df.parallax > 0]         # keep p > 0
17.         self.df       = self.df.rename(columns = \
18.                 {'vmag':'m', 'parallax':'p', 'class':'spectrum'})   # relabel columns
19.
20.         p         = self.df['p'].tolist()
21.         d         = [1/p0 for p0 in p]
22.
23.         self.df['d'] = d                                       # add d to df
24.         self.Nstars  = len(self.df)
25.
26.     def build_spectral_type_temperature_dictionary(self):
27.         st_list = ['O','B','A','F','G','K','M']
28.         st_t_limits = {}
29.         st_t_limits['O'] = [33000,40000]   # arbitrary upper lim.
30.         st_t_limits['B'] = [10000,33000]
31.         st_t_limits['A'] = [7300, 10000]
32.         st_t_limits['F'] = [6000, 7300 ]
33.         st_t_limits['G'] = [5300, 6000 ]
34.         st_t_limits['K'] = [3900, 5300 ]
35.         st_t_limits['M'] = [2300, 10000]
36.
37.         for s in st_list:
38.             [t0,t1] = st_t_limits[s]
39.             for i in range(0,10):
40.                 sub_class = s+str(i)           # e.g., 'G' + '3' --> 'G3'
41.                 temp = t1 - (t1 - t0)*i/10
42.                 self.sp_temps[sub_class] = int(temp)
43.
44.     def get_star_spectral_type_and_class(self):
45.         self.sp = self.df['spectrum'].tolist()
46.
47.         slc  = []                              # star luminosity class
48.         sscl = []                              # star subclass
49.         for i in range(len(self.sp)):
50.             sp              = self.sp[i]       # e.g., 'G8III'
51.             star_sub_class  = sp[5:7]          # e.g., 'G8'
52.             star_lum_class  = sp[7:9]          # e.g.,    'III'
53.
54.             sscl.append(star_sub_class)
55.             slc.append(star_lum_class)
56.
57.         self.df['lum_class']   = slc
58.         self.df['sp_subclass'] = sscl
59.
60.     def get_MLT_values(self):                  # get abs. mag, luminosity, tem
61.         m      = self.df['m'].tolist()
62.         d      = self.df['d'].tolist()
63.         lc     = self.df['lum_class'].tolist()
64.         spsub  = self.df['sp_subclass'].tolist()
65.
66.         star_M =[]; star_T =[]; star_L =[];
67.
68.         for i in range(0,self.Nstars):
69.             M       = m[i] - 5*math.log10(d[i]/10)
70.
71.             subclass   = spsub[i]              # e.g., 'G8'
72.             lum_class  = lc[i]                 # e.g., 'III'
73.             if lum_class != '':
74.                 T = self.sp_temps[subclass]
75.                 L = (10**(4.83-M)/2.5)
```

FIGURE 2.11  Class BSC.

*(Continued)*

```
76.              else:
77.                  T  = L = 0
78.
79.              star_M.append(M)
80.              star_L.append(L)
81.              star_T.append(T)
82.
83.          print ('length of M is: ', len(star_M))
84.          self.df['M']  = star_M
85.          self.df['L']  = star_L
86.          self.df['T']  = star_T
87.
88.
89.
90.      def get_stars_in_lum_class(self,lclass):
91.          dflc = self.df.loc[self.df['lum_class'] == lclass]
92.          return dflc
93.
94.      def make_m_vs_T_diagram(self):
95.          mydpi=120
96.          fig = plt.figure(figsize=(1200/mydpi,1000/mydpi),dpi=mydpi)
97.          for i in ['I','II','III','IV','V']:
98.              dflc = self.get_stars_in_lum_class(i)
99.              T = dflc['T'].tolist()
100.             y = dflc['m'].tolist()
101.             x = [math.log10(t) for t in T]
102.             color = self.col_map[i]

103.             mrkr  = self.mrk_map[i]
104.             plt.scatter(x,y,s = 10, c = color, marker = mrkr,label=i)
105.
106.         ax = plt.gca()
107.         ax.set_xlim([3.4,4.4])
108.         ax.invert_xaxis()
109.         ax.invert_yaxis()
110.         plt.title("Vis. Mag. vs log10(T)")
111.         plt.legend()
112.         plt.show()
113.         plt.ylabel('V mag.')
114.         plt.xlabel('Log(T)')
115.         plt.savefig('./Fig 2.9.jpg',dpi = mydpi)
116.
117.     def make_HR_Diagram(self):
118.         mydpi=120
119.         fig = plt.figure(figsize=(1200/mydpi,1000/mydpi),dpi=mydpi)
120.         for i in ['I','II','III','IV','V']:           # add one lum. cl. at a time
121.             dflc = self.get_stars_in_lum_class(i)
122.
123.             L = dflc['L'].tolist()
124.             T = dflc['T'].tolist()
125.             x = [math.log10(t) for t in T]
126.             y = [math.log10(l) for l in L]
127.             color = self.col_map[i]
128.             mrkr  = self.mrk_map[i]
129.             plt.scatter(x,y,s = 10, c = color, marker = mrkr, label=i)
130.             print(len(dflc),i, mrkr, color)
131.
132.         ax = plt.gca()
133.         ax.set_xlim([3.4,4.4])
134.         ax.invert_xaxis()
135.
136.         plt.title("Log10(L). vs log10(T)")
137.         plt.legend()
138.         plt.ylabel('Log(L)')
139.         plt.xlabel('Log(T)')
140.         plt.show()
141.         plt.savefig('./Fig 2.10.jpg',dpi = mydpi)
142.
143. if __name__ == '__main__':
144.     bsc = BSC()
145.     bsc.build_spectral_type_temperature_dictionary()
146.     bsc.read_bsc()
147.     bsc.get_star_spectral_type_and_class()
148.     bsc.get_MLT_values()
149.     bsc.make_HR_Diagram()
150.     bsc.make_m_vs_T_diagram()
151.
```

FIGURE 2.11 (CONTINUED)    Class BSC.

The **get_star_spectral_type_and_class** function parses the input spectrum string information into new columns for subclass (i.e., B5) and luminosity class.

Plots were created by adding in one luminosity class at a time (see lines 97 and 120), selected using the **get_stars_in_lum_class** function, because this offered more flexibility in selecting colors and marker symbols to emphasize the different classes. Note, when specifying markers, the size parameter in the **plt.scatter** functions was set to 10; going much smaller tends to turn all marker shapes into dots. The functions to generate the specific plots were **make_HR_Diagram** and **make_m_vs_T_diagram**.

The class is contained in the file **bsc.py**, and running this file will execute the example code after the line 'if __name__ == '__main__'' (see line 143) and produce the above plots.

## DEBCat: FINDING THE MASS-LUMINOSITY RELATIONSHIP

Astrophysicists have long recognized that one of the most important factors determining a star's evolution is its mass, and one of the best ways of directly determining mass is through studying how binary stars orbit each other, such as from eclipsing binary studies. We will use the DEBCat catalog which contains mass, radius, temperature, and luminosity for a collection of binary systems to support our explorations. The catalog and its history are described on John Southworth's page (University of Keele, UK) at https://www.astro.keele.ac.uk/jkt/debcat/ as follows:

'**DEBCat** is a catalogue of the physical properties of well-studied detached eclipsing binaries. It was originally based on the list given by Andersen (1991A&ARv...3...91), and is updated whenever revised results are published for new eclipsing binaries or for ones already in the catalogue.

DEBCat is described in a poster presented at the Kopal conference in Litomyšl, Czech Republic, September 2014. This was written up as a conference proceedings which can be found on the NASA ADS service (2015 ASPC.496.164S) and in preprint form on the arXiv server (arXiv.1411.1219).'

There is a link (https://www.astro.keele.ac.uk/jkt/debcat/debs.dat) on the page to an ASCII copy of the data which can be downloaded by using your browser's 'Save As' feature into a '.dat' file. The downloaded file can now be converted into a csv format file (e.g., debcat.csv) using Excel (see Figure 2.12).

The data column names are easily understood; there are columns about the binary system themselves (e.g., names, periods, references), but we are

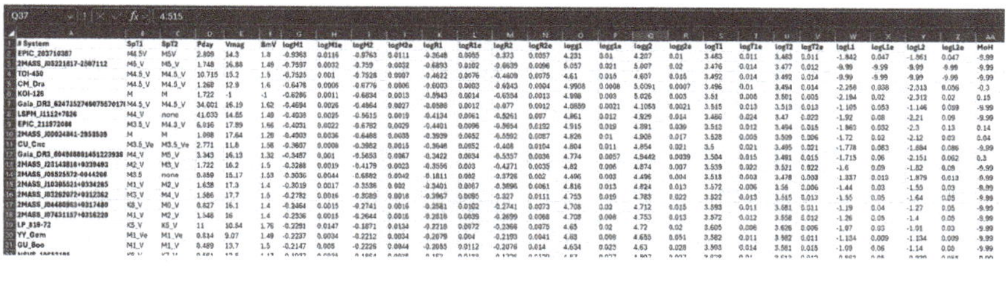

FIGURE 2.12 A section of the DEBCat data as shown by Excel.

interested in properties of the individual stars. Physical properties are generally of a form like '**logM1**' or '**logL2**,' in the first instance, the log of the first star's mass, and in the second instance, the log of the 2nd star's luminosity. Some column headers end with an 'e' (e.g., **log2Le**) to indicate the measurement error.

There are about 313 binaries in the catalog, which means there are more than 600 individual stars with L, M, T, and R measurements available to use. It is important to note that this catalog is built from the efforts of astronomers taking the time to study how each binary system behaves over time, and then undertaking some spectroscopic analysis or estimate, and measuring star sizes and masses; a very laborious and time-consuming effort, which is why it is small compared to other catalogs with hundreds of millions of entries Also, because much of the particular information must be derived from different kinds of observations, there are gaps that have yet to be filled, which means not all entries can be used; but on the other hand, the catalog is constantly growing and improving with time.

Since we are interested in individual star properties, we do not need star properties grouped by binary membership and will avoid that burden by merging data imported into our program so we will add column **logL2** to the end of column **logL1** and simply refer to it as **logL**, and similarly for **logM**, **logR**, **logT**, and spectral type (**SpT**).

DEBCat therefore gives us access to a collection of fundamental physical properties: L, M, R, and T, along with an assigned spectral type. We know that a sphere of surface temperature T and radius R will have a luminosity L given by:

$$L = 4\pi R^2 \sigma T^4 \tag{2.8}$$

That is, the area of the sphere times the energy released by each square meter at temperature T. How do our stars match up with this very simple model? How does star mass affect temperature? Luminosity? Does radius change with temperature? These are questions we can now explore using this data, but first, it is important to remember that astronomers have identified different types of stars, different luminosity classes (I–V) which show whether they are on the Main Sequence or are giants or supergiants. We will differentiate among these luminosity classes using colors and markers in our plots. (Do not be confused by this choice of color coding scheme which is intended to differentiate among the different luminosity classes; in reality, there will be stars of different colors from red to blue among class V type stars even though class V stars are drawn using blue dots in Figures 2.13–2.15).

Class **DEBCat** was developed to explore relationships among star properties. In Figure 2.13 we see a plot of **logL** against **logT** – this is essentially an HR diagram and we see how the different luminosity classes are well separated. We can readily see the Main Sequence (class V) has a slope of

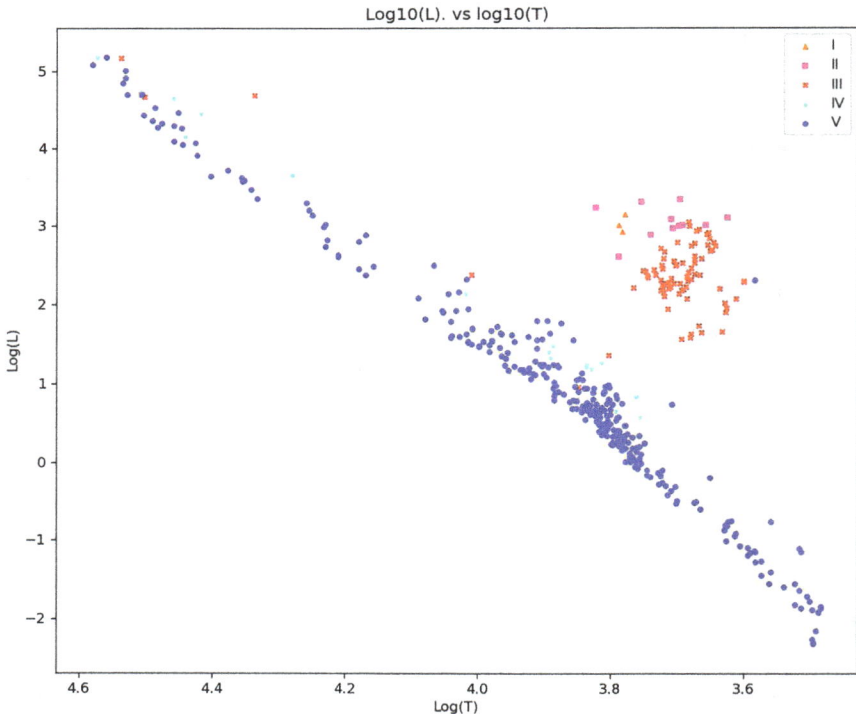

FIGURE 2.13 The HR diagram. The Main Sequence stars appear well differentiated from the giants and supergiants. The slope here is easily seen to be about 6, which means the radius must be growing almost linearly with temperature for Main Sequence stars.

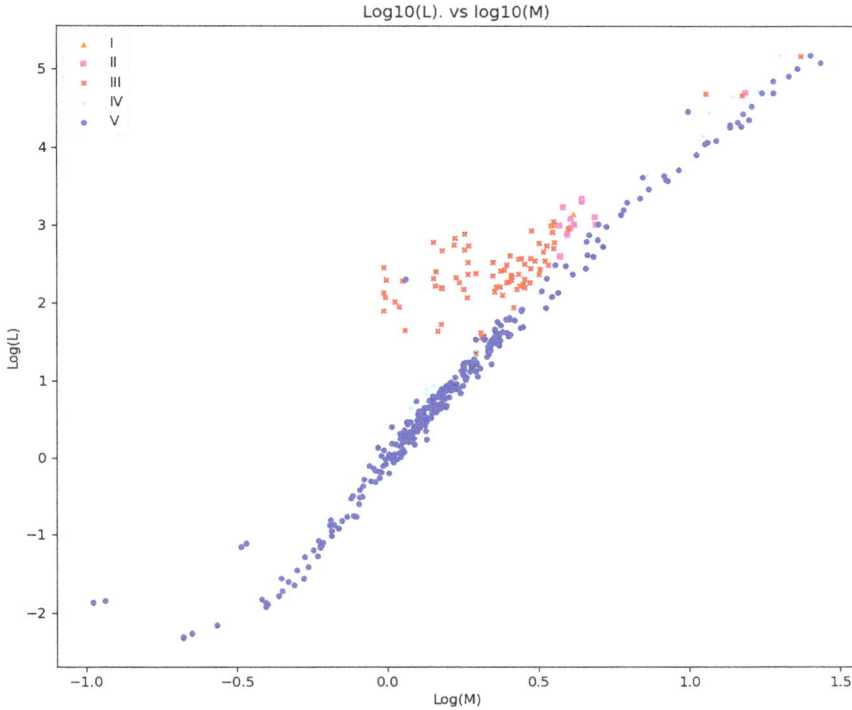

FIGURE 2.14    Plotting log(L) against log(M) shows a clear and well-defined relationship for the Main Sequence stars.

about 6 which means that for these stars, L ~ T$^6$, and based on Equation 2.8, suggests the radius is growing linearly with temperature on the Main Sequence.

How does L depend on M? A plot of **logL** against **logM** is shown in Figure 2.14 and its easily seen slope of 3.5 demonstrates another wonderful relationship, the famous L ~ M$^{3.5}$ luminosity-mass rule for Main Sequence stars.

And finally, we can explore how star radius depends on temperature by plotting **logR** against **logT** as shown in Figure 2.15, and we see that the star size (R) grows linearly with surface temperature (T) for Main Sequence stars.

With four different fundamental properties, there are 12 pairs of interactions to explore and the DEBCat data is a great starting point for student and classroom demonstrations in basic astrophysics that justify our confidence in using a simple geometric model like a sphere at a specified temperature, as a starting point in modelling a star's luminosity.

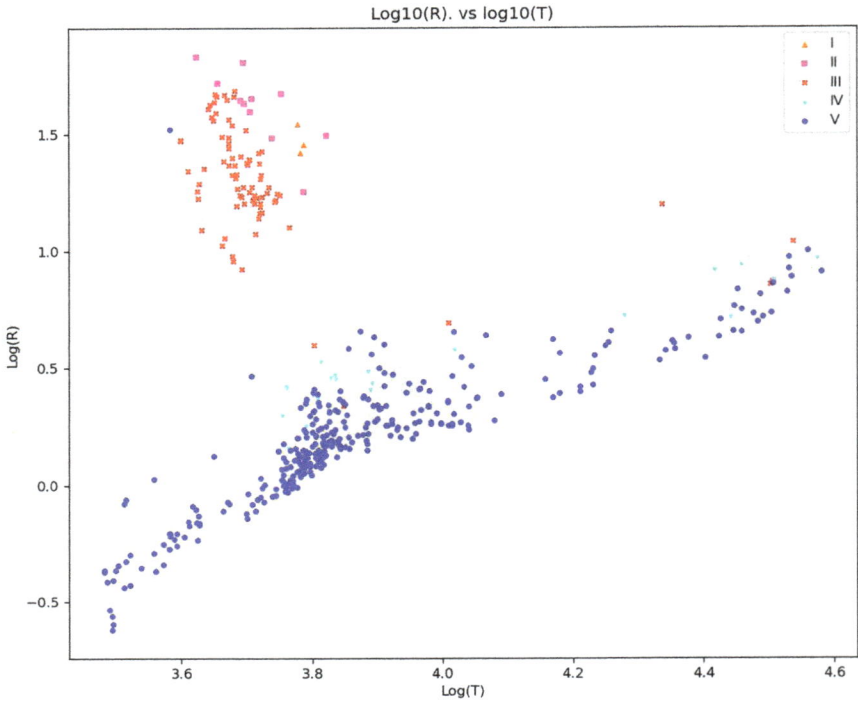

FIGURE 2.15 A plot of log(R) against log(T) shows Main Sequence stars grow almost linearly with temperature.

## CLASS DEBCat PROGRAMMING NOTES

Class DEBCat (shown in Figure 2.16) has a very similar structure to class K3L but with a few interesting and important differences. First, a new working dataframe is created by stacking columns together (lines 20–24), and second, a tricky string parsing problem had to be solved. Because we needed to know the various luminosity class types, we needed to extract this information from DEBCat's spectrum information columns SpT1 and SpT2. Because of the nature of the problem, we are using data that is often incomplete and multi-sourced, the spectrum description strings could vary widely with possible forms like G, G8, M3.5, M3_V, M3.5_V, G1_IV-V, A7_Vm, etc. This range of possible formats reflects the fact that the catalog is assembled from results and notations used by different researchers' contributions. The general trend is the spectra descriptions are of the form XY.Y_Zm, where X is the main spectral type (O, B, A,...) and Y.Y is a number showing the subtype; and there is an underline separator, followed by a roman numeral Z, and possibly some additional characters. Very often it might simply be Y instead of Y.Y for the subclass – or it is missing altogether;

```
1.  import pandas as pd
2.  import matplotlib.pyplot as plt
3.  import math
4.  import re
5.
6.
7.  class DEBCAT():
8.      def __init__(self):
9.
10.         self.sp_temps = {}
11.         self.df       = pd.DataFrame()
12.         self.col_map  = {'I':'tab:orange', 'II':'tab:pink', 'III': 'r','IV':'cyan','V':'b'}
13.         self.mrk_map  = {'I':"^", 'II':",", 'III': "x",'IV':"1",'V':"o"}
14.         self.read_debcat()
15.         self.get_star_spectral_type_and_class()
16.
17.
18.     def read_debcat(self):
19.         df0    = pd.read_csv('debcat.csv')
20.         self.df['logM'] = df0['logM1'].tolist() + df0['logM2'].tolist()
21.         self.df['logT'] = df0['logT1'].tolist() + df0['logT2'].tolist()
22.         self.df['logL'] = df0['logL1'].tolist() + df0['logL2'].tolist()
23.         self.df['Sp']   = df0['SpT1'].tolist()  + df0['SpT2'].tolist()
24.         self.df['logR'] = df0['logR1'].tolist() + df0['logR2'].tolist()
25.
26.         self.df        = self.df[self.df['Sp'] != 'none']
27.         self.df        = self.df[self.df['Sp'] != 'none']
28.
29.         self.Nstars    = len(self.df)
30.
31.
32.     def get_star_spectral_type_and_class(self):
33.         sp       = self.df['Sp'].tolist()
34.         sp_lc    = []                               # star luminosity class
35.         sp_subcl = []                               # star subclass
36.         for i in range(len(sp)):
37.             spstr = sp[i]
38.             self.res=re.findall(r'(?is)([A-Z])([O0-9.]{0,3})([_]{0,1})(IV|V?I{0,3})(.*)',spstr)
39.             [(spc,sub,symb,lc,buf)]=self.res
40.             sp_subcl.append(spc+sub)
41.             sp_lc.append(lc)
42.
43.         self.df['Sp_lum_class']    = sp_lc
44.         self.df['Sp_subclass']     = sp_subcl
45.
46.
47.     def make_HR_Diagram(self):
48.         mydpi=120
49.         fig = plt.figure(figsize=(1200/mydpi,1000/mydpi),dpi=mydpi)
50.         df0 = self.df[self.df['logL'] > -9.0]
51.
52.         for i in ['I','II','III','IV','V']:
53.             df1   = df0.loc[df0['Sp_lum_class'] == i]
54.             y     = df1['logL'].tolist()
55.             x     = df1['logT'].tolist()
56.             color = self.col_map[i]
57.             mrkr  = self.mrk_map[i]
58.             plt.scatter(x,y,s = 10, c = color, marker = mrkr,label=i)
59.         ax = plt.gca()
60.         ax.invert_xaxis()
61.         plt.title("Log10(L). vs log10(T)")
62.         plt.legend()
63.         plt.xlabel('Log(T)')
64.         plt.ylabel('Log(L)')
65.         plt.show()
66.         plt.savefig('./Fig 2.13.jpg',dpi = mydpi)
67.
68.     def make_logL_vs_logM_Diagram(self):
69.         mydpi=120
70.         fig = plt.figure(figsize=(1200/mydpi,1000/mydpi),dpi=mydpi)
71.         df0 = self.df[self.df['logL'] > -9.0]
72.         for i in ['I','II','III','IV','V']:
73.             df1   = df0.loc[df0['Sp_lum_class'] == i]
74.             y     = df1['logL'].tolist()
75.             x     = df1['logM'].tolist()
```

FIGURE 2.16   Class DEBCAT.                                          (*Continued*)

```
76.                 color = self.col_map[i]
77.                 mrkr  = self.mrk_map[i]
78.                 plt.scatter(x,y,s = 10, c = color, marker = mrkr,label=i)
79.             plt.title("Log10(L). vs log10(M)")
80.             plt.legend()
81.             plt.xlabel('Log(M)')
82.             plt.ylabel('Log(L)')
83.             plt.show()
84.             plt.savefig('./Fig 2.14.jpg',dpi = mydpi)
85.
86.        def make_logR_vs_logT_Diagram(self):
87.             mydpi=120
88.             fig = plt.figure(figsize=(1200/mydpi,1000/mydpi),dpi=mydpi)
89.             df0 = self.df[self.df['logL'] > -9.0]
90.             for i in ['I','II','III','IV','V']:
91.                 df1   = df0.loc[df0['Sp_lum_class'] == i]
92.                 y     = df1['logR'].tolist()
93.                 x     = df1['logT'].tolist()
94.                 color = self.col_map[i]
95.                 mrkr  = self.mrk_map[i]
96.                 plt.scatter(x,y,s = 10, c = color, marker = mrkr,label=i)
97.             plt.title("Log10(R). vs log10(T)")
98.             plt.legend()
99.             plt.xlabel('Log(T)')
100.            plt.ylabel('Log(R)')
101.            plt.show()
102.            plt.savefig('./Fig 2.15.jpg',dpi = mydpi)
103.
104.
105.    if __name__ == '__main__':
106.        dbc = DEBCAT()
107.
108.        dbc.make_HR_Diagram()
109.        dbc.make_logL_vs_logM_Diagram()
110.        dbc.make_logR_vs_logT_Diagram()
111.
```

FIGURE 2.16 (CONTINUED)   Class DEBCAT.

sometimes the separator is missing; sometimes the roman numeral Z is missing. And sometimes, an additional element is included like 'e' or 'm,' or a range of values. There are a small number of entries with even other forms, but for the sake of brevity, we will ignore these.

Our problem then is of how are we to extract luminosity class data from these spectrum description strings? Our solution was to use the Python **Re** library's **findall()** function with regex string matching specifiers as follows:

$$\text{re.findall}\Big(r'(?is)\big(\big[A-Z\big]\big)\big(\big[O0-9.\big]\{0,3\}\big)\big(\big[\_\big]\{0,1\}\big)$$

$$\big(IV \mid V?I\{0,3\}\big)\big(.*\big)',spstr\Big)$$

The key parts here are the () terms shown in boldface which attempt to breakdown descriptors in a format like XY.Y_Zm into five character groups(X)(Y.Y)(_)(Z)(m).

([A-Z]): this matches a single uppercase letter for the spectral classes like O, B, A…

([O0-9.]{0,3}): matches groups of numbers and decimal points; there might be 0 or 3 of them (the curly brackets adds this capability), so this set matches subclasses 3, 3.5, or even a missing specifier. We included an 'O' (uppercase letter 'o') since we noticed a typo in one instance where 'o' was used instead of zero – it is easy for a human curator to make errors like this or to copy an original source error, and so this is a useful work around.

([_]{0,1}) Here the matching allows for the presence (or absence) of an underscore character. Again, some descriptors include it; others not.

(IV|V?I{0,3}) This is the part that searches for the luminosity class roman numerals. It reads as follows: accept either IV, or V, or 0-3 I characters.

(.*) This catches any remaining character.

The command **re.findall()**, with these matching specifiers returns 5 parameters (see line 39) and the luminosity class is contained in the fourth.

As with the **K3L** class, plots are created by adding stars by luminosity class which allows us to assign markers and colors by class.

For color choices, beyond the simple single-letter base color specifiers (e.g., 'r' for 'red'), other color options can be used such as tableau ('tab. orange') – see line 12.

Note also entries without spectral type information were excluded since luminosity was of prime interest, but this restriction could be usefully relaxed. Finally, if other relationships are of interest, then a more general plotting function of the form 'plot_logX_vs_logY' could be developed based on the existing ones, instead of our customized solutions for logL vs logT, etc.

## SUMMARY

In this chapter, we saw how powerful and compelling decent graphical representations of data can be for identifying relationships among different variables, and for processes that have power-law behaviors, using log-log plots can demonstrate their presence and magnitude. A more advanced approach would use curve fitting techniques to find best match models, but we were still able to reveal behaviors and underlying astrophysical rules governing fundamental star properties. We also saw how to download and import catalogs into dataframes, and then separate out the variables of interest. Because different researchers used slightly different notations when categorizing stars, we found the character pattern matching re.findall function could be used to parse the different notations to produce a consistent one for our use.

In the next chapter, we will consider the problem of how to detect periodic signals embedded in data and show how the Fast Fourier Transform can be used to identify such signals so essential data can be succinctly summarized.

## CHAPTER 3

# Signals and Trends

S OME OF THE MOST important tools available to researchers are mathematical transformations that can reshape data into a different conceptual form that reveals important features that are hard to quantify otherwise. For example, a data sequence might contain a sine wave. The sequence might have thousands of numbers, but they all might obey a very simple rule from trigonometry and be readily summarized by as little as three numbers: amplitude, frequency (or period), and phase. And just as a mean (average) and standard deviation can powerfully summarize a dataset, or, as we saw in the last chapter, a power-law (Kepler's third law) can summarize essential relationships for a million asteroid orbits, knowing the sine-wave properties can summarize data sequences involving waves. A Fourier Transform (FT) is a mathematical tool that can operate on such a sequence and reveal the underlying (hopefully small) set of parameters; the FT transforms the data from a sequence of varying values into the world of associated parameter combinations.

At first this might not seem very extraordinary, but what happens when there is background noise confusing the wave's appearance? Or if there are multiple waves? Under such circumstances, a FT can extract underlying wave parameters. The mathematics behind the FT is quite complicated, and the calculations are very intensive and beyond the scope of this work. There is a highly efficient version called the Fast Fourier Transform (FFT) that can be very effective, but it has some important constraints such as requiring input data to be evenly spaced and the number of data points must be a power of 2, for example, 256, 1024, 4096, etc. Our goals will be to introduce the reader to FTs by showing how they can be applied to simple

DOI: 10.1201/9781003600046-3

models and learn how they function and then to more complicated real-world weather-related dataset. Our approach will show code based on Python libraries, and we will encounter common data manipulation issues and concerns.

However, a word of warning is in order; when working with complex tools like the FFT, the results can be very rewarding to explore and lead to great success in data interpretation or great failure. There is a fine line between having reasonable and justified confidence in the results and becoming seduced by false results from noise. Taking advanced course-work, consulting with more skilled practitioners and crosschecking against models are essential to avoid chasing noise signals – think of the TV shows with ghost hunters using overly sensitive scientific/engineering detectors who gasp with each electronic device fluctuation and think they are measuring something real. To prevent researchers from following spurious results, scientists will often devise double-blind experiments or inject test signals to see if they are properly identified. For example, when trying to detect gravity waves, physicists with the LIGO experiment were searching for very weak signals in their data. Under such circumstances, unknown to the researchers actually analyzing the raw data, false signals were deliberately injected by team leaders into the data every now and then to test whether the researchers could detect them and to crosscheck against erroneous reports and false positives.

Because the FFT is such a powerful tool, in this chapter we will explore using it and applying it to a meteorological dataset from the Irish Government's Met Office (www.met.ie) to see what patterns are present. But first we will create test sequence-based sine waves to show how the FFT can extract wave properties, even in the presence of multiple waves and noise. Our task therefore consists of the following elements: Create test sequences, use an FFT solution to analyze them, and then apply our techniques to real-world weather data.

## TESTING THE FFT

Fortunately, for us, the hard work of writing the code for a FFT has already been done, and we will use the NUMPY library version. The NUMPY FFT function does require some wrapper code so we can match our data to its requirements. To see how all this works, we will need some easily understood samples, test signals we will create, and see how the FFT processes them. Class **fft_demo** was developed to meet these needs and combines tools to create test sequences, calculates the FFT, and plots the results.

The FFT output is an array of values associated with an array of frequencies, and we are mainly interested in seeing where the peaks/maxima are, so we learn which frequencies are the most important. For simplicity, we will only use the Power Spectrum, in which only the magnitude of the FFT at a given frequency is used and phase information ignored.

Since the intent here is to see if the FFT function can recover/reveal/detect wave amplitudes and frequencies/periods even when degraded by noise let's look at the results from a pair of models where the noise standard deviation is 0.1 and 3 times the wave amplitude respectively (see Figures 3.1 and 3.2).

In Figure 3.1, the sine wave clearly dominates, is easily detected and revealed by the power spectrum. In our scenarios, we used a period of **p** = 48 steps. Because our FFT function only operates on an input array, it is up to the code developer to provide the interpretation of what units of time and frequency are in use. In our models, when we invoked a scenario, we also specified a timestep size of 1/24, so in real-time units, the period is 1/24 * 48 = 2. Hence the frequency is 1/**p** = 0.5, and indeed our power spectra show a peak there. (We chose 1/24 as the step size instead of 1, to both demonstrate how to use the step size parameter, and also because

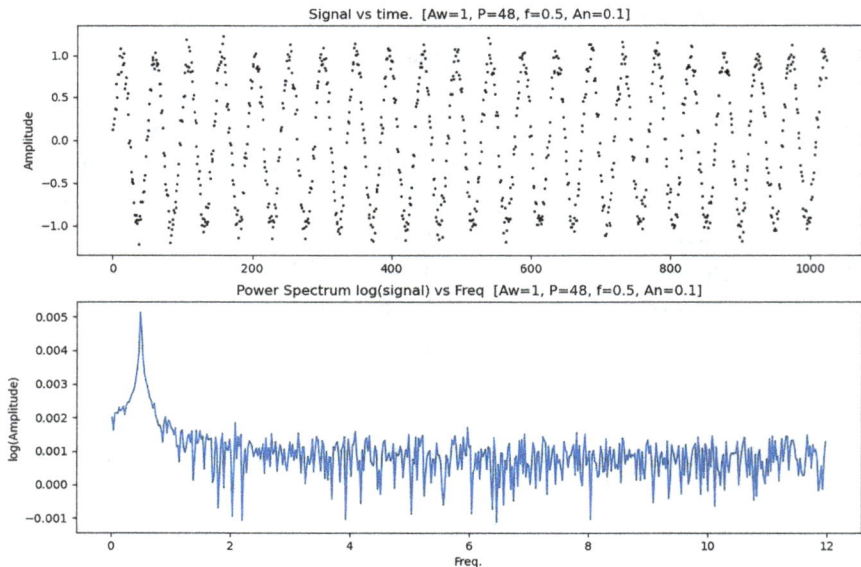

FIGURE 3.1   The FFT of a signal where the wave amplitude was ten times greater than the noise standard deviation. The sine wave component is very easily seen (top), and the power spectrum shows a peak at f = 0.5 (bottom).

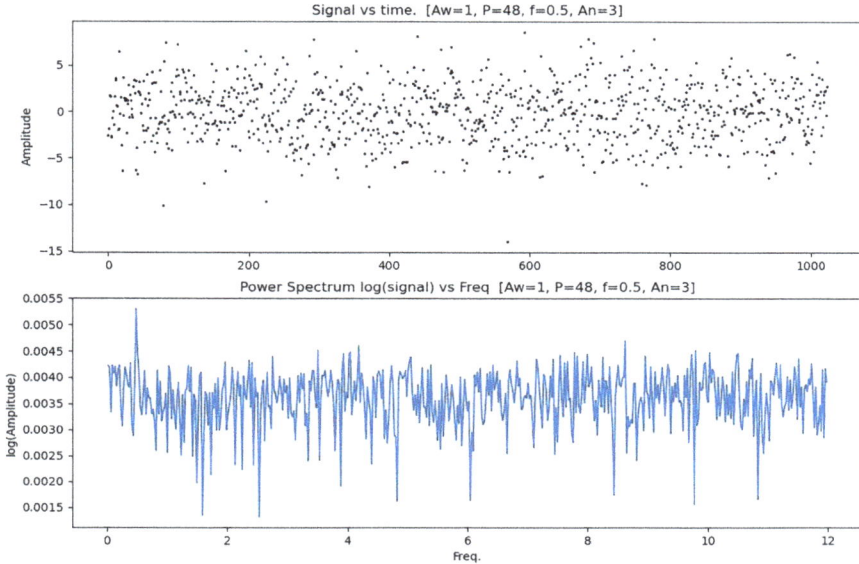

FIGURE 3.2  Making the noise standard deviation three times the wave amplitude degrades the signal so the sine wave is very difficult to see visually (top), but the FFT still shows a peak at f=0.5.

when we will look at the www.met.ie data, the measurements we use will be hourly, so a step size of 1/24 produces frequencies of 1/day.)

In producing our charts, we used a log scaling on the y-axis because very often there can be a very wide range of spectrum peak heights, and the log scaling compresses the y-scale.

In Figure 3.3, the charts from Figure 3.2 are recreated using a linear scaling, and the peak at 0.5 is very clear. The lesson here is important: small peaks in log plots can be very significant!

There is no right answer as to whether linear or log scaling should be used. It depends on what other peaks of interest are present and their relative heights and on what the user is trying to communicate, so it is up to the user to decide, perhaps informed by considering choices other specialists made.

So our models have been successful in demonstrating how an FFT can 'pull' information from noisy data, and they also serve to give a sense of how researchers can find it very exciting to examine a signal and discover previously unknown or unexpected information hiding in the noise.

And this brings us to the very important technique of signal injection. In our models, we combined a wave with noise, and we could say that we

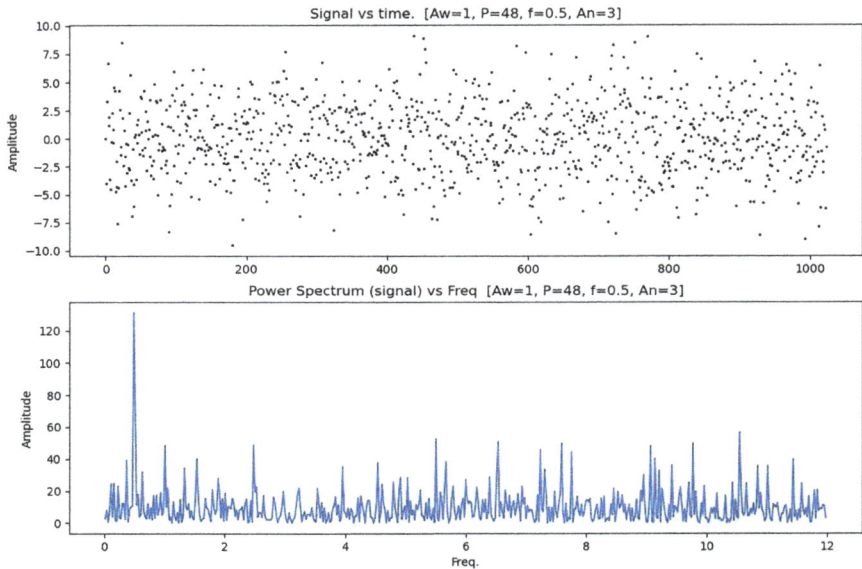

FIGURE 3.3    The same results as from those used in Figure 3.2 but without the log scaling on the y-axis.

injected a wave signal into the noise signal. With other data, we might want to deliberately inject a test signal for a few reasons. First, to cross-check whether our computer code is working properly; that indeed we used the correct time-step and scaling, so for example, an injected wave of period **p** actually produces a peak at **f**=1/**p**; if not, there is something wrong with the code. Second, by injecting a wave, we could adjust its amplitude to match a data peak of interest, which would allow us to at least make a statement that the data peak was consistent with a signal of amplitude **A** and period **p**. Third, real-world data might have many sampling problems such as gaps in the data that might create confusing and false peaks in the power spectra. Deliberately injecting a signal near the frequencies/peaks of interest can provide a useful sense of how an ideal signal is represented in the power spectrum near those frequencies – especially if the injected signal is created with the same sampling limitations (gaps).

Our code was therefore designed to facilitate exploring user-created data signals using FFT techniques and will be incorporated into the later code to be used when exploring weather data by creating a new class (**met_ie**) as a child of the **fft_demo** class so it will inherit its functions. This again shows the power of Object-Oriented Design, where the code used to solve one problem can be used in code intended to solve a different one. We will now discuss the details of the **fft_demo** class.

## CLASS fft_demo PROGRAMMING NOTES

The complete code for **fft_demo** is shown in Figure 3.4 and we will now review some of the design considerations.

Class **fft_demo** uses a strategy where a global array called **self.signal** is maintained and into this we can add as many signal components as we want such as multiple sine waves and noise to build a test signal.

```python
1.  import numpy as np
2.  import math
3.  import matplotlib.pyplot as plt
4.
5.  class fft_demo:
6.
7.      def __init__(self):
8.          pass
9.
10.     def add_test_signal(self,A,p,N):
11.         signal0 = [0]*N
12.         if A > 0:
13.             if p == 0:
14.                 signal0 = np.random.normal(0,A,N)
15.             else:
16.                 pi = math.pi
17.                 signal0 = list(range(N))
18.                 signal0 = [A*math.sin(2*pi*s/p) for s in signal0]
19.         signal0     = np.array(signal0)
20.         self.signal = self.signal + signal0
21.
22.     def plot_data(self,title_str,vals):
23.         x = list(range(len(vals)))
24.         plt.scatter(x,vals,marker='.', s=6, c='k')
25.         plt.title(title_str)
26.
27.
28.     def find_largest_pow2_subset(self, N):
29.         n       = 2
30.         while 2**n <= N:
31.             n += 1
32.         N2      = 2**(n-1)
33.         return N2
34.
35.     def get_fft(self,yvals, time_step, dstr, mode):
36.
37.
38.         ps          = np.abs(np.fft.fft(yvals))**2
39.
40.         if mode == 'log':
41.             ps1     = np.array([math.log10(p) for p in ps])
42.             ps = ps1
43.         ps = ps/len(ps)
44.         freqs       = np.fft.fftfreq(yvals.size, time_step)
45.         N3 = int(len(yvals)/2)
46.         idx =       np.array(list(range(1,N3)))
47.         plt.plot(freqs[idx], ps[idx])
48.         plt.title(dstr)
49.
50.
51.     def run_test_scenarios(self,N,Aw, P, An, time_step,mode):
52.         self.signal  = np.zeros(N)
53.         self.add_test_signal(Aw,P,N)        # add ampl. Aw wave, period P
54.         self.add_test_signal(An,0,N)        # add noise of ampl. An
55.         f = 1/time_step/P
56.         fstr=', f='+str(f)
57.         sig_str = '  [Aw=' + str(Aw)+ ', P=' + str(P)+fstr+', An='+str(An)+']'
58.         mydpi=100
59.         fig = plt.figure(figsize=(1200/mydpi,800/mydpi),dpi=mydpi)
60.
61.         plt.subplot(2,1,1)
62.         self.plot_data('Signal vs time.'+sig_str,self.signal)
```

FIGURE 3.4    **Class fft_demo.**

*(Continued)*

```
63.
64.        plt.ylabel('Amplitude')
65.
66.        plt.subplot(2,1,2)
67.        plt.xlabel('Freq.')
68.        plt.ylabel('Amplitude')
69.
70.        if mode == 'log':
71.            plt.ylabel('log(Amplitude)')
72.            dstr          = 'Power Spectrum log(signal) vs Freq' + sig_str
73.        else:
74.            dstr          = 'Power Spectrum (signal) vs Freq' + sig_str
75.        N2  = self.find_largest_pow2_subset(len(self.signal))
76.
77.        self.get_fft(self.signal[-N2:], time_step, dstr,mode)
78.
79.        plt.show()
80.        plt.savefig(self.figure_jpg,dpi = mydpi)
81.
82.
83. if __name__ == '__main__':
84.     fd = fft_demo()
85.     fd.figure_jpg = './Fig 3.1.jpg'
86.     fd.run_test_scenarios(1024, 1, 48, .1, 1/24,'log')
87.     fd.figure_jpg = './Fig 3.2.jpg'
88.     fd.run_test_scenarios(1024, 1, 48,  3, 1/24,'log')
89.     fd.figure_jpg = './Fig 3.3.jpg'
90.     fd.run_test_scenarios(1024, 1, 48,  3, 1/24,'lin')
91.
```

FIGURE 3.4 (CONTINUED)   **Class fft_demo.**

It contains a function **add_test_signal(self, A, p, N)** to generate a signal array of size **N**, that could be either a sine wave or noise. If the period, **p**, is zero, the amplitude, **A**, is the standard deviation of a random noise (mean zero)-generated signal; otherwise, **A** is the amplitude of a sine wave with period **p**. For example, **self.add_test_signal(2, 24, 1024)** would create a 1024-long array containing a sine wave of amplitude 2 and period 24 array elements/steps, while **self.add_test_signal(3, 0, 1024)** would add 1024 random (Gaussian) numbers with standard deviation 3, to the 1024 long array. Because **self.signal** is a global array, the **add_test_signal** function can be called repeatedly to add additional different sine waves to the signal array.

The code is designed to run scenarios (e.g., lines 84–90) where sine wave(s) can be mixed with Gaussian noise. A scenario is run using a five-parameter command like **run_test_scenarios(1024, 1, 48, .1, 1/24, 'log')** which in this example causes an array of length 1024 to be created using a sine wave of amplitude 1, period 48, mixed with noise of standard deviation 0.1. Each array step is 1/24 of a time unit, and log-scaling is used for the y-axis (e.g., line 86).

Function **run_test_scenarios()** can invoke the **add_test_signal** function multiple times to build up complex signal components, which are added to **self.signal** (line 20) and create a two-panel display output (lines 58–80).

```
1. np.fft.fftfreq(8,1)
2. Out[80]: array([ 0.   ,  0.125,  0.25 ,  0.375, -0.5  , -0.375, -0.25 , -0.125])
```

FIGURE 3.5  Running the **np.fft.fftfreq** function for an array of length 8 produces an array of frequency values reflecting symmetries in the underlying calculation. Only the first half are needed for the power spectrum.

Running a scenario passes the generated signal to the **get_fft()** function which produces the power spectrum. The required parameters are the input array, the timestep size, a descriptive string, and the selected mode (use 'log' for log-scaling on the y-axis).

To find the largest data subset that meets the power-of-two FFT data size requirement, in function **find_largest_pow2_subset()**, powers of two are tested against the array size to find the largest (lines 28–33) so any input length array can be used.

Function **get_fft()** calculates the FFT at line 38 and the properly scaled frequency array at line 44.

On line 43, a simple rescaling is done (dividing by N) because the power spectrum can increase with array size, and our datasets can be very large.

Two-panel plots are created using the **plt.subplot(r,c,n)** command (lines 59 and 61), where **r** is the number of rows, **c** is the number of columns, and **n** is the subplot count, so the first plot (of the raw data) is referenced using arguments (2,1,1) and the second (of the power spectrum) by (2,1,2). We use the **plt.tight_layout()** feature to create appropriate spacing around the subplots (line 67).

The reader is encouraged to explore the indexing used by NUMPY's FFT libraries. For example, **np.fft.fftfreq(8,1)** returns the array shown in Figure 3.5, line 2. For our power spectra, we only need the first half of the frequency array. This is achieved by creating a list of indexes, half the size of the frequency array (see Figure 3.4 lines 45–46). Note the indexes start from 1 instead of 0, to avoid the f=0 spike that can occur in a power spectrum if the mean of the data is not zero. Just as problematic is when the mean() is zero, since then the power spectrum is zero at f=0, and problematic for log scaling since log(0) is undefined.

## EXPLORING METEOROLOGICAL DATA

FFTs are very useful for detecting patterns in data series, and longer the data series is, the better its ability to detect long period (low frequency) effects, which suggests they could be usefully applied to long-term weather data collections. In this section, we will use a dataset from the Irish government, downloadable from www.met.ie, since it has high-resolution (hourly) data going back almost 80 years. We will use data from recent years, from a weather station called 'Valentia Island' to illustrate our analysis. (I selected this dataset for exploration because I have always wondered about Ireland's very unpredictable weather and wanted to see what patterns, if any, could be detected.)

The data can be downloaded by going to the https://www.met.ie/climate/available-data/historical-data web page. On the left side, you can specify the data resolution (hourly/daily/monthly), select the county ('Kerry'), and then the Valentia Observatory. At this point, you are offered a list of variables to choose from using control clicks. There is also a link 'Download the full data series' and if you click on this link, three more links appear at the page bottom, one of which 'Download the full hourly data series' will initiate a download of all the Valentia data – regardless of which selections you made. The downloaded file will have to be extracted from a zip file and is now available for use.

There is a problem though, the hourly data for about 20 variables, spanning 80 years is large – more than 50 MB. This is unwieldy and unsuitable for many users and so we need to make the data more manageable. (For anyone using major databases, this kind of problem arises regularly, and so the solution we show here could be adapted to many other scenarios.) Our solution was to divide the data into a collection of annual files using class **split_by_year** which we will now describe.

## Class split_by_year Programming Notes

Class **split_by_year** is a short Python script (see Figure 3.6) that reads in a csv file with date information and splits the data into different files by year, for example, 'valentia2021.csv' in a local directory. It is a useful example of how to process a large file, and based on pattern matching done on each line, write the line to an appropriate output file. While a large computer could read in the data into a dataframe, smaller computers might not be able to process enormous datasets all at once.

The data file to be subdivided is passed as an argument during instantiation (line 36). The base name is saved (it doesn't matter if, for example, 'valentia.csv' or 'valentia' is used). Note when splitting the name string using **split('.')**, a list is returned, and the base name will be the first element.

It uses the **os** library to test whether a local directory 'YEARSCSV' exists, and if not, creates it (lines 10–11).

```
 1. import os
 2.
 3. class split_by_year:
 4.
 5.     def __init__(self,fname):
 6.         self.basename = fname.split('.')[0]
 7.         self.fname = fname
 8.         self.years_dir = './YEARSCSV/'
 9.
10.         if(not os.path.exists(self.years_dir)):
11.             os.makedirs(self.years_dir)
12.
13.     def split_csv_by_year(self):
14.         fout = ''
15.         active_yr = 0
16.         count = 0
17.         for line in open(self.fname):
18.             csv_row = line.split()
19.             if count == 23:                    # skipping info text
20.                 headers=line                   # headers on line 23
21.             elif count > 23:
22.                 [m,d,y] = csv_row[0].split('/')
23.
24.                 if y != active_yr:
25.                     if fout:
26.                         fout.close()
27.                     active_yr = y
28.                     outname = self.years_dir+self.basename+str(y)+'.csv'
29.                     fout = open(outname,'w')
30.                     fout.write(headers)
31.                 fout.write(line)
32.             count += 1
33.         fout.close()
34.
35. if __name__ == '__main__':
36.     md = split_by_year('valentia.csv')
37.     md.split_csv_by_year()
38.
```

FIGURE 3.6    Class **split_by_year** divides up the large raw datafile into yearly files for convenience.

The input file is opened for reading one row of data at a time and each line is stored as string **line** (see line 17).

Each row is a comma separated list of fields, but the very first field (for the 'date') has both date and time strings separated by a whitespace (e.g., '1/1/1944 2:00'). This whitespace is the only whitespace character on a line and separates the date from the rest of a comma separated list, so **line. split()** results in a list of two parts: the date and everything else (see line 18). The date information can be split using the '/' separator, allowing us to find the year (see line 22). The first 22 lines of data are ignored, and those of line 23 are used for headers (see line 20). When data for the next year is found, the previous year's output file is closed (line 26), a new output file created, and the headers are written (lines 27–30). The current line of data is written to the active output file at line 30. The script is run by creating an instance of the **split_by_year class**, and running the **split_csv_by_year()** function (see lines 36–37). After running **split_by_yr** using our 'valentia.csv' data, a local directory 'YEARSCSV' is populated with files 'valentia1944.csv,' 'valentia1945.csv'… By dividing the data into annual files, only those years of interest need to be selected by any other program using these files.

## WEATHER DATA FREQUENCIES

To explore the www.met.ie data, a new class was developed (**met_ie**) that includes a function, **display_var()**, that will display a two panel plot of a selected variable, showing the raw data in one, and its power spectrum in the other, for a specified year range, with the option of using log scaling on the spectrum. Class **met_ie** is a child of **fft_demo** and needs both the **fft_demo.py** and **met_ie.py** files so the classes can be imported. Figure 3.7 shows how it might be used, and the results are shown in Figure 3.8.

Because the data length is so long, only the last 365 days in the selected data is used for the top plot, to give an overall sense of annual behaviors. Because the rain is so persistent throughout the year, linear scaling was used because log scaling suppressed the peaks' visual impression.

```
1.    met = met_ie()
2.    met.load_df(2000,2022)
3.    met.display_FFT('rain','lin','Power',      './Fig 3.8.jpg')
```

FIGURE 3.7    Shows a simple script to display rain data from 2000–2022, using log scaling.

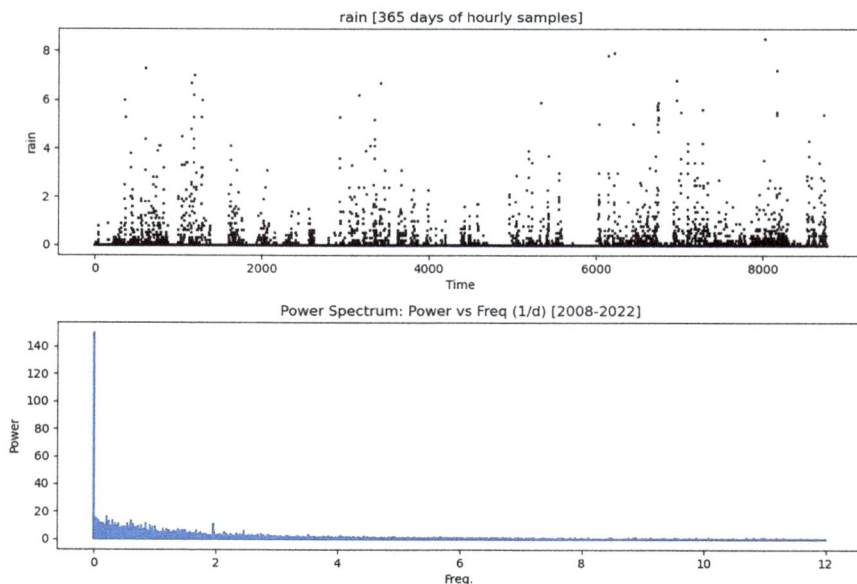

FIGURE 3.8    Charts produced by the script in Figure 3.7.

The results are very interesting. Certainly, there are periods (gaps) in the rain, but it looks like rain is a possibility at any time of the year, perhaps a little less so in the summer. The largest peaks in the spectrum correspond to an annual trend ($f \sim .003$), a twice daily cycle ($f = 2$). Interpreting other peaks is best left to meteorologists, but we can at least also note there is also a broad distribution of energy over a wide range of frequencies, a characteristic of turbulence/chaos.

What about temperature? The results are dramatic and are shown in Figure 3.9.

A word of caution. Because the annual trend is so strong, it dominates the linear scaling. If a log scaling is used, a set of peaks appears at frequencies of 1, 2, 3, and 4 times/day. (See Figure 3.10). Without further study, these most likely include artifacts. A good test would be to inject a very strong signal of period 1/4 days and see if it also produces similar resonances, which is a likely test done by a researcher. Another test would be to isolate the background level by applying a smoothing (averaging) filter and then subtracting the background from the original data and seeing what signals remained.

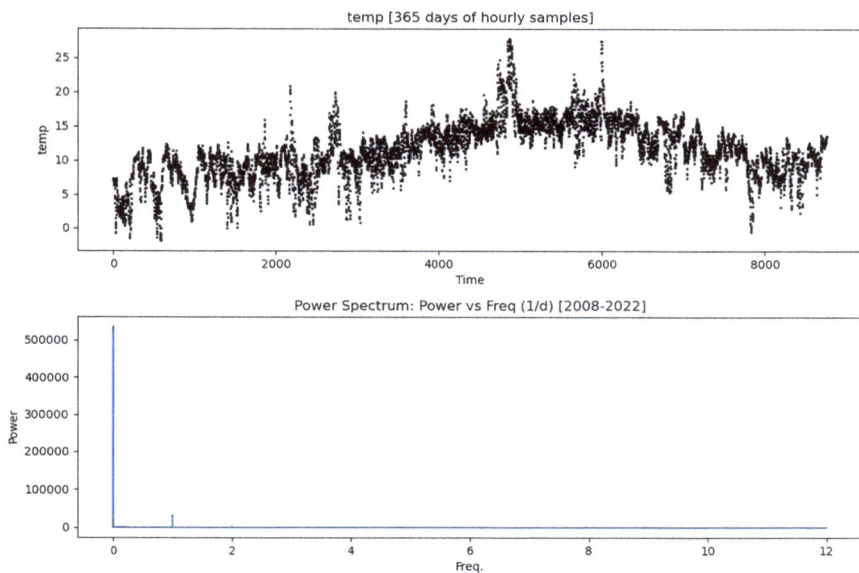

FIGURE 3.9 Results from 2015–2022 temperature measurements. The top panel shows a strong annual cycle, and the bottom panel shows two dominant frequencies, annual and daily.

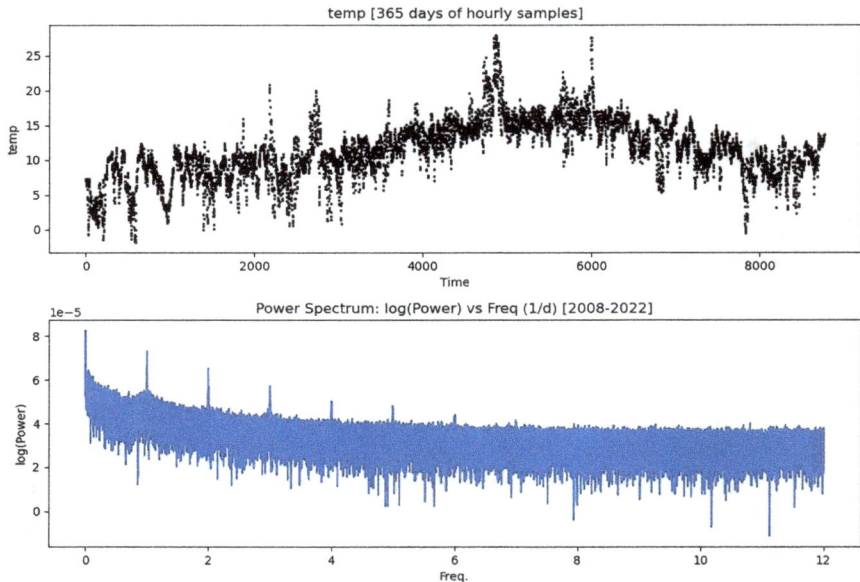

FIGURE 3.10  Plotting the spectrum from Figure 3.9 using log scaling shows a pattern of signals at multiples of 1/d, which are probably artifacts of the very strong annual signal. In any case, a broad contribution at all scales is easily seen, again indicating the presence of chaos.

## WEATHER DATA TRENDS

Because the www.met.ie data spans almost 80 years, it warrants testing to see if there are climate change trends present, and the results are shown for the hourly temperature measurements in Figure 3.11. The chart was produced by instantiating the **met_ie** class and then running the **display_trend()** function (see Figure 3.12). The chart plots a linear fit to the smoothed data, and the slope is 0.006195 degrees Celsius per year.

To build the chart and detect the trend, the data was first smoothed by applying an averaging filter (window) of size 365*24 to span a year's worth of data, so in the smoothed data array, the value at position i is the average of all the data between **i-365*24** and **i** which suppressed daily and seasonal effects. This asymmetric averaging was done to simplify finding a linear fit, that is, allowing it to be applied in the range [**365*24, N**].

With the smoothed data, the Numpy polynomial fitting routine could now be applied and looks like: **c1, c2 = np.polyfit(xvals, yvals, n)**. In **polyfit**, **n** is the order of the polynomial (we used **n** = 1 to obtain a linear fit), and the function returned the slope (**c1**) and intercept (**c0**). Using **n** > 1 would produce higher order polynomial fits and a longer list of coefficients.

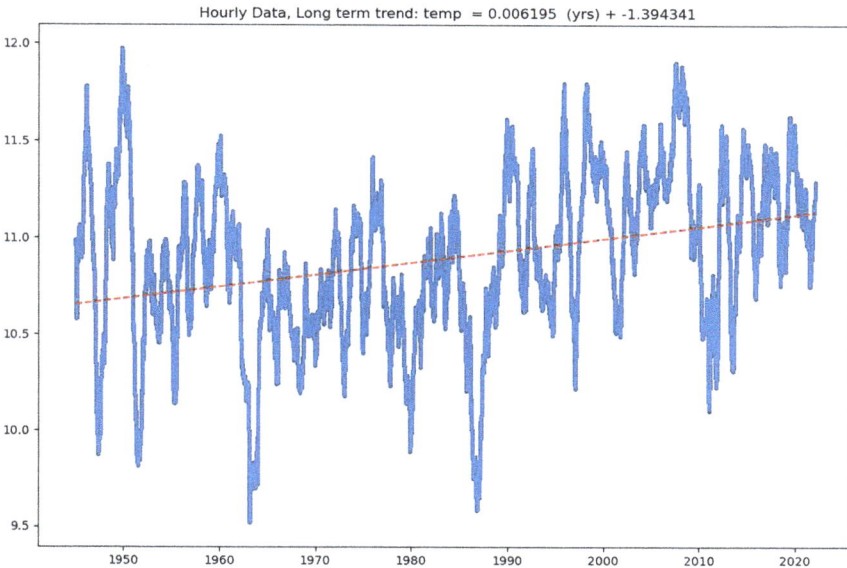

Hourly Data, Long term trend: temp = 0.006195 (yrs) + -1.394341

FIGURE 3.11    The long-term trends in the hourly temperature measurements show an annual growth rate of 0.006195 degrees Celsius per year between 1944 and 2022.

```
1.     met.load_df(1944,2022)
2.     met.display_trend('temp', 24*365)
3.
```

FIGURE 3.12    Using the **met_ie** class to find trends in hourly temperatures between 1944 and 2022, with a smoothing window of 1 year (365*24).

## CLASS met_ie PROGRAMMING NOTES

The **met_ie** class (see Figure 3.13) used to access the Valentia weather station data found at www.met.ie which we reformatted as a collection of annual files because it contains hourly measurements spanning almost 80 years. The data is stored in local directory (see line 10). Data is read into a dataframe **self.df** for analysis and manipulation. Function **load_df()** allows the user to select a year range for analysis (see lines 26, 120, and 125). The data contains about 20 columns/variables, and only 4 columns are selected for input (see lines 24–25). The data comes from the www. met.ie site (line 9), and data is read for all specified years (line 13) using **get_site_data()**, which reads in each year's data (line 17) and concatenates it to the **self.df** dataframe (line 18). To avoid problems with dataframe indexing that can happen with concatenation, the index is rebuilt each time new data is added to the dataframe (line 19). To support future development, a **daynum** variable was added so measurements could be

```
 1. import pandas as pd
 2. import numpy as np
 3. import matplotlib.pyplot as plt
 4. from fft_demo import fft_demo
 5.
 6. class met_ie(fft_demo):
 7.
 8.     def __init__(self):
 9.         self.site              = 'valentia'
10.         self.site_root_dir     = './MET_IE_CSV/'
11.         self.df_new = pd.DataFrame()
12.
13.     def get_site_data(self,site, yr_list):
14.         yr_list.sort()
15.         for yr in yr_list:
16.             print(site, str(yr))
17.             self.read_site(site, yr)
18.             self.df = pd.concat([self.df,self.df_new])   #stack data
19.             self.df.index = range(len(self.df.index))
20.
21.     def read_site(self, site, year):
22.         filename=self.site_root_dir+site+str(year)+'.csv'
23.         print(filename)
24.         self.df_new = pd.read_csv(filename, delimiter=',', \
25.                       usecols=['rain','temp','msl','wdsp'])
26.     def load_df(self,ymin, ymax):
27.         self.df = pd.DataFrame()
28.         self.ymin = ymin
29.         self.ymax = ymax
30.         self.yrs               = '['+str(ymin)+'-'+str(ymax)+']'
31.
32.         yrlist                 = list(range(ymin,ymax))
33.         self.get_site_data(self.site,yrlist)
34.         nrows                  = len(self.df)
35.         daynum                 = list(range(0,nrows))
36.         daynum                 = [d/24. for d in daynum]
37.         self.df['daynum']      = daynum               # decimal day number
38.
39.         self.N2 = self.find_largest_pow2_subset(nrows)
40.         nyrs = int(self.N2/365/24)
41.         ymin = ymax - nyrs
42.         self.yrs_recent        = '['+str(ymin)+'-'+str(ymax)+']'
43.
44.
45.     def display_FFT(self,colname,mode,colstr,figname):
46.         df0 = pd.DataFrame()
47.         avg = self.df[colname].mean()
48.         df0[colname] = self.df[colname].fillna(avg)      # replace nan with avg
49.
50.         N = len(df0)
51.         self.signal = np.zeros(N)
52.         #self.add_test_signal(5,2*24,N)                   # uncomment to add tracer signal
53.         vals = np.array(df0[colname]) + self.signal
54.         timestep = 1/24
55.
56.         mydpi=100
57.         fig = plt.figure(figsize=(1200/mydpi,800/mydpi),dpi=mydpi)
58.         plt.subplot(2,1,1)
59.         plt.xlabel('Time')
60.         plt.ylabel(colname)
61.         self.plot_data(colname+ ' [365 days of hourly samples]',vals[-365*24:])
62.         plt.subplot(2,1,2)
63.         plt.xlabel('Freq.')
64.         plt.ylabel(colstr)
65.         dstr   ='Power Spectrum: '+colstr+' vs Freq (1/d) '+ self.yrs_recent
66.         self.get_fft(vals[-self.N2:],timestep,dstr,mode)
67.         plt.subplots_adjust(hspace=.3)
68.         plt.show()
69.         plt.savefig(figname,dpi = mydpi)
70.
71.
72.     def apply_df_smoothing(self,varname, window_size):
73.         indata = self.df[varname]
74.         N = len(indata)
75.         sm = np.zeros(N)
76.         tenpct = int(N/10)
```

FIGURE 3.13    Class met_ie.                                                                 (*Continued*)

```
 77.              for i in range(N):
 78.                  if i%tenpct == 0:
 79.                      print('doing line  ',i)
 80.                  if i < window_size:
 81.                      sm[i] = 0
 82.                  else:
 83.                      wlen =window_size
 84.                      subset = indata[i-window_size:i]
 85.                      sm[i] = subset.sum()/wlen
 86.
 87.              self.df[varname+'_sm'] = sm
 88.
 89.      def display_trend(self,varname,wsize):
 90.
 91.              met.apply_df_smoothing(varname,wsize)
 92.              met.df[varname+'_flat'] = met.df[varname] - met.df[varname+'_sm'] # unused
 93.
 94.              yvals = met.df[varname+'_sm'].tolist()
 95.              N = len(yvals) - wsize
 96.              yvals = yvals[-N:]
 97.              time = np.array(list(range(N)))
 98.              time = time/24
 99.              time = time/365                        # elapsed time is now in yrs
100.              time = time + self.ymin + wsize/24/365
101.
102.              mydpi=100
103.              fig = plt.figure(figsize=(1200/mydpi,800/mydpi),dpi=mydpi)
104.              plt.scatter(time,yvals,s=4)
105.
106.              c1,c0 = np.polyfit(time, yvals, 1)           # add trend line
107.              y = [c0+c1*time[0], c0+c1*time[N-1]]
108.              x = [time[0],time[N-1]]
109.              plt.plot(x,y,color='r',linestyle='dashed')
110.
111.              pstr = varname +'  = '+str(round(c1,6)) +'  ' + ' (yrs) + '+str(round(c0,6))
112.              plt.title(' Hourly Data, Long term trend: '+ pstr)
113.              print(c0, c1)
114.              plt.show()
115.              plt.savefig('./Fig 3.11.jpg')
116.
117. if __name__ == '__main__':
118.
119.      met = met_ie()
120.      met.load_df(2000,2022)
121.      met.display_FFT('rain','lin','Power',      './Fig 3.8.jpg')
122.      met.display_FFT('temp','lin','Power',      './Fig 3.9.jpg')
123.      met.display_FFT('temp','log','log(Power)','./Fig 3.10.jpg')
124.
125.      met.load_df(1944,2022)
126.      met.display_trend('temp', 24*365)
```

FIGURE 3.13 (CONTINUED)    Class met_ie.

referred to using decimal days (lines 35–37). Once the class is instantiated (e.g., line 119), any column can be specified for analysis, over the desired time-range, using the **display_FFT()** or **display_trend()** functions. At line 52, a capability to add a tracer signal can be used by uncommenting the line. The FFT is found using the **get_fft** function from the parent class **fft_demo** and titles are added (e.g., line 60). At line 48, the missing values are replaced with the column's data average. Titles are generated at lines 30 and 42, with the **yrs_recent** string used for the FFT which, being limited to a power of two, can use significantly less than the data's requested time span. When working with trends for a variable name like '**rain**,' the smoothed version produced by the **apply_df_smoothing()** function is called 'rain_sm,' and a flattened version called 'rain_flat.' And similarly for other variable names.

Because the dataset is large, it can take a few minutes to do the numerically intensive smoothing operations, and so a current active line number is printed to the console at 10% intervals as feedback to the user the process is continuing (see lines 78–79). The coefficients for the linear fit are found at line 106 using **np.polyfit**, and the trendline constructed at lines 105–109.

## SUMMARY

In this chapter, we saw how a powerful tool like the FFT can reveal and help document patterns and signals embedded in data collected over long periods of time. Accessing such data can require parsing it and dividing it into more manageable pieces, so we also showed how this could be done with CSV files.

With simple demonstrations and simulated signals, we saw that the FFT can detect surprisingly weak signals, ones that would be hard to detect with the eye. An ability to create pure signals was seen to be very useful to verify that the code was correctly assigning frequencies and also for deliberately injecting signals of known frequency to see if artifacts are being created. Applied to a set of Irish weather reports, it revealed annual and daily patterns, and, not surprisingly, data such as rain measurements, are consistent with turbulence and chaotic processes; processes that can manifest a continuum of signals/energy across a wide range of frequencies.

We also explored how long-term trends can be detected and documented by using curve fitting tools such as **np.polyfit**; while we used a linear model, higher order models are easily done. We found evidence the weather station location increased in temperature by about 0.5 degrees Celsius since 1944. There is much more that could be done with the weather data, and it would be easy to build on our code for further studies. In the next chapter, we will use simulations to explore the gravitational effects of various mass distributions, to verify some important results from calculus and show how more complex systems can be analyzed if they can be assembled from simpler ones.

## CHAPTER 4

# Gravity Fields and Mass Distributions

I N THIS CHAPTER, WE will explore gravitational effects for different kinds of mass distributions (geometries or shapes). Spherically symmetric solids and shells are of particular interest since there are some very powerful results from calculus related to them, which we can verify using numerical methods. Not all distributions have solutions that are easily found by merely applying calculus techniques. Many particle distribution problems are also relevant to other research areas such as electrostatics, where charged flat surfaces and rings are important and so some of these geometries will also be studied.

Most of us should be familiar with Newton's law of gravity (see Equation 4.1) which tells us the magnitude of the force of attraction F between two masses (m and M) separated by distance r. G is the universal gravitational constant. G takes on different values depending on what units of measure you use, metric, imperial, and so on. Its role is to set the scale of the force, consistent with the units. Both m and M experience the same size of force pulling them together, and traditionally, we introduce the equation by stating m and M are 'point masses,' which means the masses are treated as being point like. If m and M represent the masses of a person and the Earth, respectively, then F would be their weight at the Earth's surface – we would set r equal to the Earth's radius. In fact, we could rewrite the law as:

$$F = m\frac{GM}{r^2} = mg \tag{4.1}$$

DOI: 10.1201/9781003600046-4

where g is 9.8 m/s/s in scientific (MKSA) units (when we insert the Earth's mass as M and the Earth's radius for r), the acceleration due to gravity at the Earth's surface. (In this form, we see F = ma, as described in elementary dynamics.)

But neither the person nor the Earth are point-like, so why does Newton's law of gravity still apply? The answer is, when matter is distributed uniformly in shells and spheres, such as in the Earth where there are shells of different densities surrounding the spherical core, each shell, as seen from the outside, behaves as if all the shell's matter was concentrated at the center, and when inside the shell, the shells effect disappears. These remarkable and beautiful results can be proven using calculus. The Earth is close to being spherical, and using the law of gravity, and treating the Earth as a point mass allowed scientists to predict satellite orbital speeds. In fact, during the early days of the Space Age, scientists had amateur astronomers submit timing measurements of when satellites would pass, as a cross check of their models, and they discovered a systematic discrepancy in their predictions that was subsequently explained by updating their understanding of the Earth's shape, eventually characterizing it as being slightly asymmetric and pear-shaped.

Interestingly, the electric force between charged particles also has a form similar to Newton's law of gravity: Coulomb's law states the force between two particles with charges q and Q, separated by a distance r, obeys the rule:

$$F = k\frac{qQ}{r^2} \tag{4.2}$$

The scaling constant is k and is much larger than G, which means electrical forces tend to be stronger than gravitational ones. Unlike for mass which is always positive, q and Q could be either positive or negative. Therefore, while masses always and relentlessly try to pull themselves together, electrical charges with the same polarity (i.e., both are positive, or both are negative) will repel each other; opposites will of course attract.

Both Coulomb's and Newton's laws here are called inverse-square laws because they have $r^2$ in the denominator. The power of two is not an approximation to some real-world value like 1.999999999. It could be argued that one way to interpret Newton's law of gravity is to say there is something (M) causing an equal effect in all directions and that effect passes through a spherical surface of area $4\pi r^2$, and how concentrated the effect is at a distance r is the central effect diluted by the area, and since a sphere's surface area depends precisely on a power of 2, the dilution must depend on a power of 2.

In our notation, we emphasize the concept of using a smaller test mass (m) being influenced by a larger mass (M). We could just as easily have written $m_1$ and $m_2$. But the idea of using a small mass is often important for two reasons. First, being small, its own effect minimally disturbs that of the larger mass, and so it can be used to study the larger mass' gravitational effects. Second, very often we will use a unit test mass (m = 1), and the resulting forces it experiences from being moved near the larger mass tell us about the gravitational field – in physics, a force field tells us what the effect would be on a unit mass (or a unit charge for electrostatics). It's similar in a way to buying food at a grocery store, where you pay attention to the price per pound; the actual price is the price per unit times the quantity to be purchased. If a unit mass experiences a force of 8, a mass of 5 would experience a force of 40 at that same position. Put another way, when observing the test mass, the gravitational field characterizes the effects of the larger mass being studied, while the gravitational force tells us about the effect on the test mass by the larger.

Because gravity is always attractive, it tends to pull things together, and this is why stars and planets tend to be round and objects can get trapped in orbits. However, in spite of its apparent simplicity, like other inverse-square laws, it contains a mathematical problem – as r approaches zero, the force becomes infinite. This means that when we create models to study, we do not want to 'step' on a point mass when sampling!

In the rest of this chapter, we will develop models to explore issues like the following:

- How gravity disappears inside a spherical shell, so as to verify a result from calculus?

- How, outside a shell, gravity behaves like all the mass is concentrated at the center?

- The gravitational field along any 3-D radial axis from a ring of matter in the X-Y plane

- The gravitational field along any 3-D radial axis from a disk of matter in the X-Y plane

Our interest in the disk and the ring distributions arises from the fact that they are distributions of interest in electrostatics, and also because in astronomy, many galaxies, planets, and protostars, have disk-like structures. From calculus, we learn that on the outside, the gravitational field of a uniform spherical shell is the same as if all the shell's mass was concentrated at the center. Furthermore, when passing through the shell, the gravitational force disappears. Since a uniform solid sphere is simply a collection of shells, we can now say that outside a ball, the gravitational field acts as if all the ball's mass was at a point at the center, but interior to the ball, at any point, the exterior shell has zero net effect, and only the mass interior to that point's radius counts, and it also behaves as if it is concentrated at the center.

One question we will explore for all these models is to what extent do their gravitational fields behave like that of a shell or a sphere? Do their gravitational fields obtained from adding in the effects of all points match the ideal's (all interior/enclosed mass is at the center)? To answer this, we include the ideal plots which are generated as if only the mass interior to a sampling point was relevant and that interior mass was concentrated at the center.

Our strategy will be to write functions to distribute particles according to the geometry we are interested in, which will result in a dataset consisting of the x, y, and z positions for each particle. Then, for a set of positions along a specified radial-axis of the system, we will calculate the distance (s) from each position to each particle, and then the total gravitational force is calculated at each sampling position. On completion, we will have a set of net gravitational effects for each sampled position: we can then plot the field strength versus the radial-axis position.

We will specify an observer location for each model that will define the radial being sampled and which will allow our models to be used not just for the normal axes of symmetry typically used in textbooks, but also others that we can explore.

Our simulation software system consists of six files:

- gsims.py to manage and invoke the desired model type. It imports the class definitions from the individual model class files.

- grav_sim.py holds the class definition for the parent class (grav_sim).

- class_ring.py holds the definition for the ring_model subclass.

- class_disk.py holds the definition for the disk_model subclass.

- clad_single_shell.py holds the definition for the single shell subclass.

- class_double_shell.py holds the definition for the double shell model subclass.

- class_sphere.py holds the definition for the sphere's subclass.

All subclasses have the same structure and contain functions to build the distribution, get the ideal radial profile, and create a two-panel plot: one showing the 3-D distribution, and the other the numerical and the ideal radial profiles.

## CONTROLLING APPLICATION: gsims.py

The controlling file, **gsims.py** is straightforward and very short (see Figure 4.1) and was used to generate this chapter's plots. It imports the files containing the class definitions for the five models we are exploring. All models require six numeric parameters: **r1**, **r2**, and **M** which specify the radius (r1) of the disk, ring, and shell, the radius (r2) of an outer second shell if used for the double-shell model (r2 is zero if not needed), and the total number of mass points used, while the last three specify the end of the sampling radial that extends out from the origin (0,0,0). The last

```
 1. from class_disk import disk_model
 2. from class_ring import ring_model
 3. from class_double_shell import double_shell_model
 4. from class_single_shell import single_shell_model
 5. from class_sphere import sphere_model
 6.
 7.
 8.
 9. #######################################################################
10. #
11. # Each model uses the same style of arguments
12. #
13. # model(r1, r2, M, Rx, Ry, Rz)
14. #
15. # r1 and M set the radius (r1) and total mass (M) of a system
16. #
17. # r2 is the radius of the 2nd shell in the double-shell model
18. # otherwise r2 = 0
19. #
20. # The sampling radial extends from (0,0,0) to (Rx,Ry,Rz)
21.
22.
23. ss = single_shell_model(1,0,500,5,0,0, './Fig. 4.4.jpg')
24. ss = single_shell_model(1,0,500,0,5,0, './Fig. 4.5.jpg')
25.
26. ds = double_shell_model(1,3,500,5,0,0, './Fig. 4.6.jpg')
27.
28. rm = ring_model(1,0,500, 5,0,0,      './Fig. 4.7.jpg')
29. rm = ring_model(1,0,500, 0,0,5,      './Fig. 4.8.jpg')
30.
31. dm = disk_model(1,0,500, 5,0,0,      './Fig. 4.9.jpg')
32. dm = disk_model(1,0,500, 0,0,5,      './Fig. 4.10.jpg')
33.
34. sp = sphere_model(1,0,2000,5,0,0,    './Fig. 4.11.jpg')
35.
```

FIGURE 4.1   File **gsims.py** is how we control and specify which model to use. It imports all needed classes and a model is selected by uncommenting its line. Here it was configured to produce plots for this chapter.

parameter is a path/filename to store the generated chart. This design is very extensible, and it would be easy to add other models such as parallel plate disks (capacitors!) or hemispheres, and so on.

## CLASS grav_sim PROGRAMMING NOTES

The parent class is defined in **grav_sim.py** (see Figure 4.2) and contains resources used by all models. It contains functions to initialize

```
 1. import math
 2.
 3. class grav_sim:
 4.
 5.     def __init__(self,r1,r2,M,Rx,Ry,Rz):
 6.
 7.         self.r1          = r1          # radius 1
 8.         self.r2          = r2          # if > 0, radius of second entity
 9.         self.M           = M           # Total Mass
10.         self.n           = M           # particles in M
11.         self.Obs         = [Rx,Ry,Rz]  # sampling radial end-point
12.         self.Nsteps      = 100          # radial points to sample
13.
14.         self.x           = []          # x, y, and z for model
15.         self.y           = []
16.         self.z           = []
17.         self.v           = []
18.         self.g           = []          # initialize radial gravity field
19.         self.dlist       = []          # radial sampling distances
20.
21.         self.dlist1      = []          # ideal curve 1
22.         self.trace1      = []
23.
24.         self.dlist2      = []          # ideal curve 2 for second shell
25.         self.trace2      = []
26.
27.     def get_radial_gravitational_field(self):   # exclude 'stepped on' points
28.         o = self.Obs
29.         d_Obs = math.sqrt(o[0]**2 + o[1]**2 + o[2]**2)
30.         onx  = o[0]/d_Obs
31.         ony  = o[1]/d_Obs
32.         onz  = o[2]/d_Obs                        # radial unit vector
33.         res = d_Obs/self.Nsteps
34.         N = len(self.x)
35.         self.dlist = [x*res +res/10 for x in range(0,self.Nsteps)]
36.         m1 = self.M/self.n
37.         print('Radial Samples: ', len(self.dlist))
38.         print('x values: ', len(self.x))
39.         print('count is: ', N)
40.         for d in self.dlist:
41.             gx = 0
42.             for n in range(0,N):
43.
44.                 dx      = d * onx                # sampling pos is [dx, dy, dz]
45.                 dy      = d * ony
46.                 dz      = d * onz
47.
48.                 print(dx, self.x[n])
49.                 sx      = dx - self.x[n]
50.                 sy      = dy - self.y[n]
51.                 sz      = dz - self.z[n]
52.                 s2      = sx**2 + sy**2 + sz**2
53.                 s       = math.sqrt(s2)
54.
55.                 if s > 4*res :                   # test if too close to point mass
56.                     sn      = [sx/s, sy/s, sz/s]  # unit vector to Obs
57.
58.                     cos_theta = sn[0]*onx + sn[1]*ony + sn[2]*onz
59.                     grav      = m1/s2             # mass = m1 = M/n
60.                     gravr     = grav*cos_theta    # force radial component
61.                     gx        = gx + gravr
62.             self.g.append(gx)
```

FIGURE 4.2    Class **grav_sim** is the parent class for the different geometries explored.

```
63.
64.
65.
66.    def get_ext_inv_r_squared_field(self, total_mass,radius ):
67.
68.        o = self.Obs
69.        d_Obs = math.sqrt(o[0]**2 + o[1]**2 + o[2]**2)
70.
71.        res = d_Obs/self.Nsteps
72.        self.dlist = [d*res for d in range(0,self.Nsteps)]
73.
74.        dl = []
75.        tr = []
76.
77.        for d in self.dlist:
78.            if d >= radius:
79.                tr.append(total_mass/d**2)
80.                dl.append(d)
81.        return dl, tr
82.
83.    def make_parameter_string(self):    # make a string like [r = 1,0, M = 500]'
84.
85.        str1 = str(self.r1)
86.        if self.r2 > 0:
87.            str2 = ', '+str(self.r2)
88.        else:
89.            str2 = ''
90.
91.        strm = str(self.M)
92.        self.pstr = '(r = ' + str1  + str2 + ', M = ' + strm + ')'
93.
```

FIGURE 4.2 (CONTINUED)    Class **grav_sim** is the parent class for the different geometries explored.

all models, calculate the radial gravitational fields of interest and the ideal ones, and methods needed to create the two-panel plots containing model results.

Measurements are taken along a radial from the origin out to a point **self.Obs**, defined by [**Rx, Ry, Rz**] which represents the tip of the sampling axis; [**x, y, z**] are the coordinates of the model's mass points; **g[]** is the calculated gravitational fields; and **dlist[]** are the sampling distances.

At line 27 (**get_radial_gravitational_field**), the gravitational effects of the models' mass distributions are calculated.

Each measurement involves selecting a position **d** along the sampling radial (line 40), and for each mass point's position **X**, finding the vector from **X** to **d** (i.e., **s = d − X**) (see lines 44–52) and using that to get the gravitational field (line 59). The dot product of the unit vectors for **s** and **d** gives the component of **g** along the sampling radial (lines 58 and 60).

Only the radial components of the fields are calculated because our plans are to explore symmetric distributions with sampling radials along either the x-, y-, or z-axis, even though we have generalized so the radial could be in any direction. Studying fields transverse to the radial sampling would require some additional code changes.

The parent class (**grav_sim**) uses six model command line arguments in its **__init__**() function – which are used by all models.

Note, on line 55, we ignore any point within a minimum distance of the sampling point to prevent divide by zero kinds of instabilities. If the mass distribution is intended to model a continuum, then ignoring the nearest one is probably okay since in any case, a continuum wouldn't have a concentrated point. The minimum distance used here was determined through trial and error.

It is important to appreciate the **get_radial_gravitational_field()** method works for any supplied distribution (the [x, y, z] lists for the mass points). It's where the bulk of the model number-crunching occurs and is properly embedded in the parent class as a resource to be used by any model.

Function **get_ext_inv_r_squared_field()** calculates the ideal field outside the main mass distribution(s), without having to get the distances between the sampling position and the individual masses since it assumes all mass is concentrated at the origin.

With the double-shell model we will explore later, we will calculate the ideal twice: once using both shell masses and also for the inner shell only. The idealized traces are stored in **trace1**[] and **trace2**[] with **dlist1**[] and **dlist2**[] being their corresponding positions.

The remaining method (line 83) offered by the parent support plotting roles is a small utility used to create a plotting label string.

## PARTICLE DISTRIBUTION MODELS

In the rest of this chapter, we will look at the results of the different geometries we chose to model, such as disks, shells, double-shells, rings, and spheres, since these kinds of geometrical structures could appear in electrostatic and gravitational systems of interest such as parallel plate capacitors, charged spheres (conducting or not) and rings, and also for star cluster and galaxy gravitational field modelling.

The single shell model is of particular interest because of the result from the calculus we have previously mentioned, namely that outside a shell, the mass behaves as if it is all at the center and inside the shell the gravitational forces cancel out. It also implies that outside a solid sphere, the mass should appear as being at the center, and at an interior point distance r from the center, only the mass enclosed within r contributes to the field, and that mass appears to be located at the center. So, if we can successfully model a shell's gravitational field, we are also effectively confirming the behavior for a solid sphere, and because we know what to expect, we can test if our core algorithms are working properly.

All model classes follow the same design:

1. There is a class definition and initialization

2. The initialization builds the model using a function like do_XXX_ system()

3. Building a system consists of:

   a. Adding mass points to the system

   b. Calculating the fields along the sampling radial axis

   c. Creating the idealized (mass at the center) data

   d. Plotting the mass distribution and gravity field plots

Because all our model class designs have a very similar structure, we will now provide a more detailed discussion for the single shell model one and be briefer with the others – shown at the chapter's end.

## CLASS single_shell_model PROGRAMMING NOTES

In Figure 4.3, we see the **single_shell_model** class definition. This class is a subclass of the parent (**grav_sim**) class because it invokes it as an argument (see line 5). Notice how while the **single_shell_model** class has its own __ **init__**() method, it uses the parent's to initialize variables used by all models (see line 8). As part of its initialization, it automatically runs the model by invoking the **do_single_shell_system()** (line 9). Note also, we did not need to instantiate the parent class; having access to its definitions was sufficient.

All our models are run when their subclasses are created and initialized, and running a model has the same procedure in all cases: add points to the desired mass distribution, get the radial gravitational fields, get the ideal fields, and build the two-panel plot.

The **self.add_points_to_shell()** function used to add points to the model uses a clever algorithm (found on the stackoverflow.com website) for distributing points over a shell, reasonably evenly separated, and provides the [**self.x, self.y, self.z**] data needed for calculating gravitational fields. (Each model class has its own customized function to build its mass distribution geometry.) The class has an appropriate wrapper to get the ideal field (line 40), and it has the instructions to assemble the two-panel plot. Note the plotting procedure use methods from the parent class, automatically available to the subclass.

```
1.  import math
2.  from grav_sim import grav_sim
3.  import matplotlib.pyplot as plt
4.
5.  class single_shell_model(grav_sim):
6.
7.      def __init__(self, r1, r2, M, Rx, Ry, Rz, fname):
8.          super().__init__(r1,r2,M,Rx,Ry,Rz)
9.          self.do_single_shell_system(r1, M, fname)
10.
11.     def do_single_shell_system(self,r1, M,fname):
12.         self.add_points_to_shell(r1)
13.         self.get_radial_gravitational_field()
14.         self.get_single_shell_ideal_traces(r1)
15.         self.plot_two_panel_single_shell_field(fname)
16.
17.     # Use fibonacci algorithm to distribute points on a sphere
18.     # from https://stackoverflow.com/questions/9600801/
19.     #        /evenly-distributing-n-points-on-a-sphere
20.
21.     def add_points_to_shell(self, shell_radius):
22.
23.         phi = math.pi * (math.sqrt(5.) - 1.)       # golden angle in radians
24.         N = self.n
25.         for i in range(N):
26.             y = shell_radius*(1 - (i / float(N - 1)) * 2)
27.             radius = math.sqrt(shell_radius**2 - y * y)    # radius at y
28.             theta = phi * i                          # golden angle incr.
29.
30.             x = math.cos(theta) * radius
31.             z = math.sin(theta) * radius
32.
33.             self.x.append(x)
34.             self.y.append(y)
35.             self.z.append(z)
36.
37.
38.     # Create the field where all the mass is at the center
39.     #
40.     def get_single_shell_ideal_traces(self,r1):
41.         self.dlist1,self.trace1 = self.get_ext_inv_r_squared_field(self.M, r1)
42.
43.
44.     def plot_two_panel_single_shell_field(self,img_filename):
45.         s = '[' + ','.join(str(x) for x in self.Obs ) +']'
46.         s = '   Sampling Vector: '+s
47.         self.make_parameter_string()
48.         model_name = "Single Shell Gravity Field: "+self.pstr + s
49.         mydpi=120
50.         fig = plt.figure(figsize=(1200/mydpi,600/mydpi),dpi=mydpi)
51.         plt.rcParams['axes.facecolor'] = 'white'
52.
53.         fig.suptitle(model_name)
54.         ax = fig.add_subplot(1,2,1,projection='3d')
55.         ax.scatter(self.x, self.y,self.z, color = 'blue',s=3)
56.         fig.add_subplot(1, 2, 2)
57.         plt.plot(self.dlist, self.g)
58.         plt.plot(self.dlist1,self.trace1)
59.         plt.xlabel("Distance from center")
60.         plt.ylabel("Radial Grav. Field")
61.         plt.subplots_adjust(wspace=.3)
62.         plt.show()
63.         plt.savefig(img_filename,dpi = mydpi)
64.
```

FIGURE 4.3   The **single_shell_model** class.

## SINGLE-SHELL MODEL RESULTS

To test our single-shell model, let's run it to generate a shell of radius of 1 and 500 particles. The radial sampling extends out along to x-axis to [5, 0, 0]. The results are shown in Figure 4.4. The 3-D plot shows the particles were indeed distributed nicely over the shell, and the gravitational

Single Shell Gravity Field: (r = 1, M = 500)  Sampling Vector: [5,0,0]

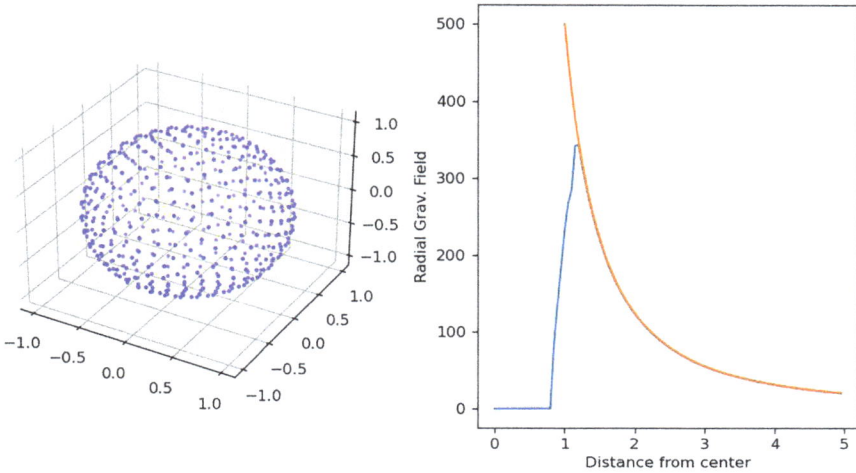

FIGURE 4.4  The results of running the single-shell model. The red measured effects match the blue ideal curve nicely.

field calculated by adding up the effects of the 500 particles (red) matches the ideal (blue) values. The gravity field disappears inside the shell as expected. The red curve does peak near r = 1, but not quite to the ideal curve's max which is $M/r^2 = 500$. This is because we avoid getting too close to any particle and system granularity.

Overall though, the model has worked very well and has demonstrated the ideal curve based on assuming the enclosed mass is concentrated at the center, really works, and also that the gravitational field is indeed zero on the inside.

Note, we did benefit from not approaching any particle too closely. If we changed the sampling axis to lie along the y-axis, a slightly different result is obtained (see Figure 4.5).

## DOUBLE-SHELL MASS DISTRIBUTIONS

The **double-shell** class is a simple modification of the single-shell class. The main differences are that the **do_double_shell()** method adds a second shell dividing the mass evenly between the two shells, generates two ideal curves, and adding a second ideal trace to the plot's gravitational field.

With a double-shell system, we expect each shell to contribute a gravity field curve like that in Figures 4.4 or 4.5. For example, if we have two

Single Shell Gravity Field: (r = 1, M = 500)   Sampling Vector: [0,5,0]

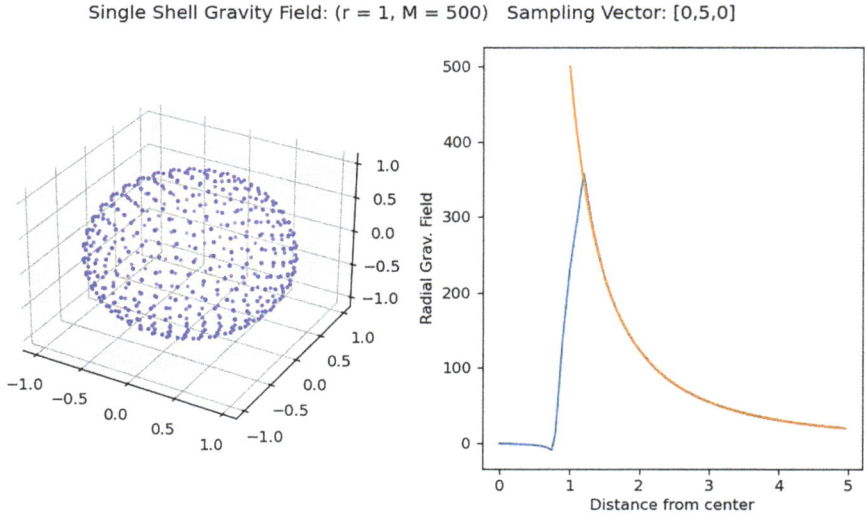

FIGURE 4.5   Sampling along the y-axis enabled us to straddle a particle that exerted excessive pulls inward and outward. Otherwise, the behavior is very close to the ideal.

shells, one at **r1** = 1 and the other at **r2** = 3, with a total mass of 500, then the inner shell would have a max of $250/1^2$, and the outer a max of $250/3^2$. Because of our previous results, being on the inside, the inner shell is unaware of the outer shell's existence, while just outside the outer shell, all mass appears concentrated at the center. Between the shells, only the inner shell's mass is a factor and appears concentrated at the center.

Running the model, like before, from **gsims.py** produces the results shown in Figure 4.6, in agreement with our expectations.

Not surprisingly, there are small sampling effect consequences near the shell boundaries, but the major conclusion is that indeed, when working with uniform shells, the results from calculus hold true.

## RING AND DISK DISTRIBUTIONS

These distributions are frequently found in electrostatics but do have some interesting applications in astronomy such as for planetary ring systems and galaxy disks. For completeness we will show the class definitions for both, with minimal discussions on them since all our models have similar coding structure.

Figures 4.7 and 4.8 show the results when exploring the ring's field along the x-axis and along the z-axis, and there are some immediate interesting

Double Shell Gravity Field: (r = 1, 3, M = 500)   Sampling Vector: [5,0,0]

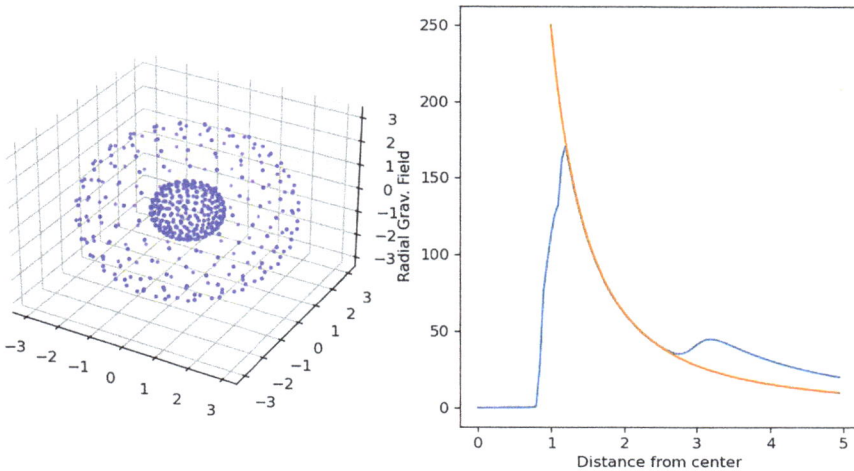

FIGURE 4.6   Double shell model results with 500 points divided between two shell of radius 1 and 3. The ideal curves (blue dots) show the 250/r² and 500/r² trends consistent with calculus.

Gravity Field for Ring: (r = 1, M = 500)   Sampling Vector: [5,0,0]

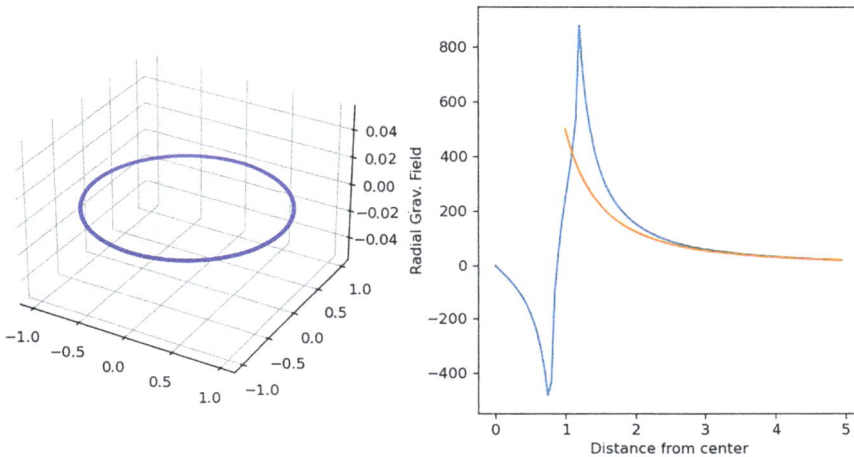

FIGURE 4.7   Ring gravity field along the x-axis.

results. First, unlike shells, the ring field does not immediately go to zero on crossing through the ring – the ring itself pulls back on the test mass. Also, not surprisingly, the field along the z-axis is attractive all the way in but goes to zero at the center. These models suggest that ideal curves based on

Gravity Field for Ring: (r = 1, M = 500)  Sampling Vector: [0,0,5]

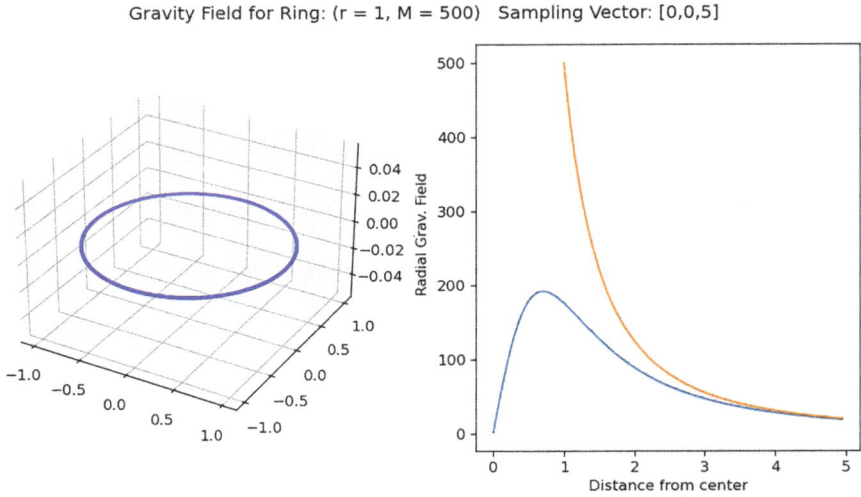

FIGURE 4.8    Ring gravity field along the z-axis.

assuming all mass appears at the center when external to a ring are only a reasonable approximation when further than about twice the ring size.

For the disk models, there were many times we sampled too close to some particles when approaching along the plane of the disk, and they had undue influence if closer than **4\*res** so points closer than this were excluded.

Gravity Field for Disk: (r = 1, M = 500)  Sampling Vector: [5,0,0]

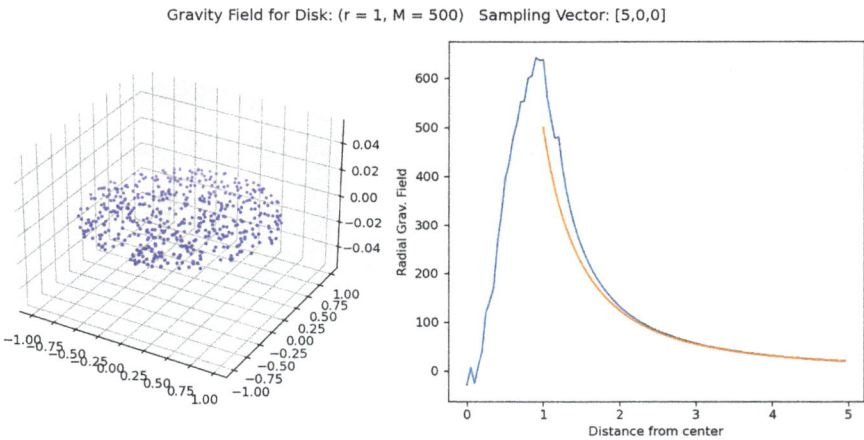

FIGURE 4.9    Approaching the disk center along the x-axis can produce noisy results from getting too close to individual particles.

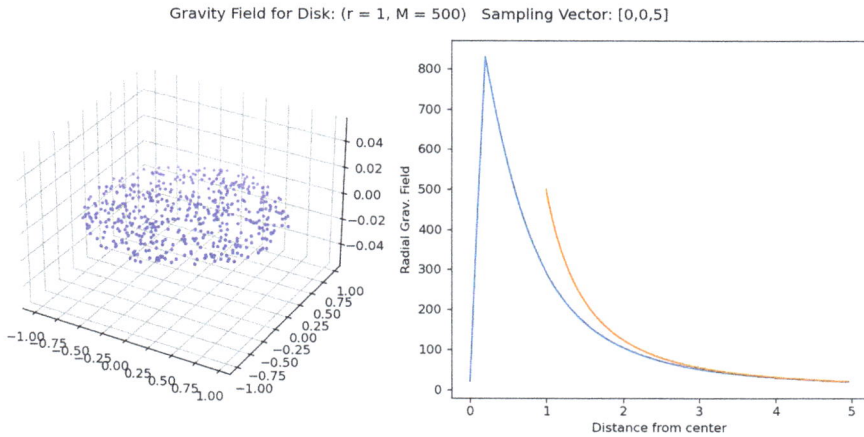

FIGURE 4.10    Disk model results. Approaching from the vertical produces smoother results.

The results of x-axis and z-axis sampling are shown in Figure 4.9. For comparison, Figure 4.10 shows the results where the sampling vector lies along the z-axis.

## SPHERICAL MASS DISTRIBUTION

The final mass distribution geometry we consider is that for a solid sphere (in our case, particles distributed throughout the sphere's volume). You should be able to predict how the gravity field should be; passing through the surface, the outer layers should cancel and only the effect of the interior enclosed mass should matter. So, at distance r, M(r), the enclosed mass should be the volume times the density and therefore scale as $r^3$ for a uniform density. However, the field is scaling as $M/r^2$ and so the overall effect is that in the interior, the field is linear with r.

This model uses class **sphere_model**. To create the point distribution, points were sampled by randomly placing points in a cube extending from +/- r in the cartesian directions and only keeping those that lay within the sphere's radius. (We could easily modify the algorithm to build a cubic distribution by removing this constraint.)

A ball model with 2000 points distributed within a sphere of radius 1, with x-axis sampling was run by using **sp = sphere_model(1, 0, 2000, 5,0, 0, fname)** in the controlling file **gsims.py** and the results are shown in Figure 4.11.

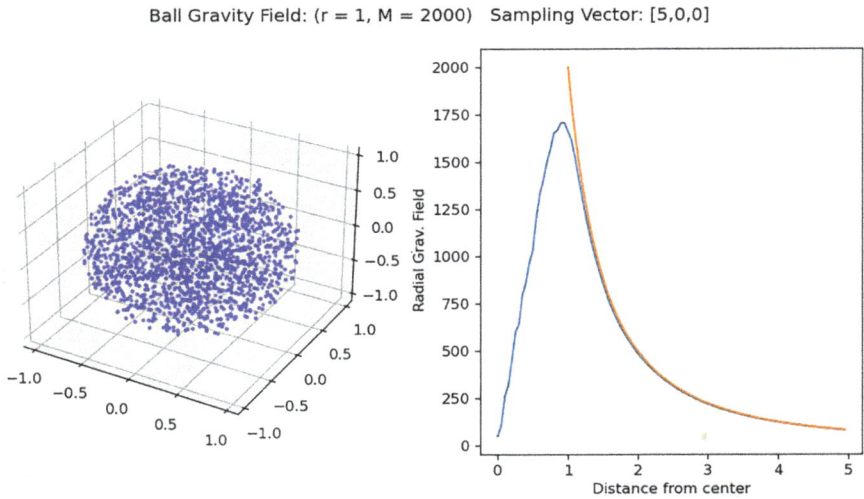

FIGURE 4.11 Gravity field for 2000 points distributed in a sphere of radius 1. On the outside, the field matches the $1/r^2$ ideal (blue dots), while in the interior, it grows linearly as expected. The results match our expectations very well since we have a $1/r^2$ (blue dots) trend outside the ball, and a linear one on the interior. This nicely demonstrates the underlying principles.

## A NOTE ABOUT ELECTROSTATICS

The models we created here assumed there was a positive attractive force at all times – this was appropriate because gravity is always attractive. For electrostatics however, opposite charges attract, but like charges repel. To modify our models to explore electrostatics, create separate classes based on the mass distributions with appropriate labeling and do the following:

1. Use q and Q instead of m and M.

2. Allow Q to be positive or negative.

3. Treat q like a test charge, which is usually considered positive.

4. Change the plot titles and labels.

5. Rename functions with names that include phrases like 'grav' or 'gravity,' etc.

6. Consider changing the sign convention when calculating force so a positive value represents a force outward. (Our models show positive fields indicating an attractive force.)

7. The spherical distribution also applies with the assumption a non-conductor is being modeled – otherwise the charges would move as far apart as they could and end up on the surface!

## PARTICLE DISTRIBUTION MODEL CODES
### Disk Model

```
 1. import random
 2. import matplotlib.pyplot as plt
 3. from grav_sim import grav_sim
 4.
 5. class disk_model(grav_sim):
 6.
 7.     def __init__(self, r1, r2, M, Rx, Ry, Rz, fname):
 8.         super().__init__(r1,r2,M,Rx,Ry,Rz)
 9.         self.do_disk_system(r1, M, fname)
10.
11.     def do_disk_system(self,r,n, fname):
12.         self.n         = n
13.         self.M         = n
14.         self.radius    = r
15.
16.         self.add_points_to_disk(r, n)
17.         self.get_radial_gravitational_field()
18.         self.get_disk_ideal_trace()
19.         self.plot_two_panel_disk_fields(fname)
20.
21.     def add_points_to_disk(self, disk_radius,M):
22.         r = disk_radius
23.         N = self.M
24.         count = 0
25.         while count < N:
26.             x = r*(random.random()*2 - 1)
27.             y = r*(random.random()*2 - 1)
28.             z = 0
29.             if (x*x + y*y ) < r*r:
30.                 self.x.append(x)
31.                 self.y.append(y)
32.                 self.z.append(z)
33.                 count = count + 1
34.
35.     def get_disk_ideal_trace(self):
36.         self.dlist1,self.trace1 = \
37.                 self.get_ext_inv_r_squared_field(self.M, self.radius)
38.
39.
40.     def plot_two_panel_disk_fields(self,img_filename):
41.         s = '[' + ','.join(str(x) for x in self.Obs ) +']'
42.         s = '   Sampling Vector: '+s
43.         self.make_parameter_string()
44.         model_name = "Gravity Field for Disk: "+self.pstr     +s
45.         mydpi=120
46.         fig = plt.figure(figsize=(1200/mydpi,600/mydpi),dpi=mydpi)
47.         plt.rcParams['axes.facecolor'] = 'white'
48.
49.         fig.suptitle(model_name)
50.         ax = fig.add_subplot(1,2,1,projection='3d')
51.         ax.scatter(self.x, self.y,self.z, color = 'blue',s=3)
52.         fig.add_subplot(1, 2, 2)
53.         plt.plot(self.dlist, self.g)
54.         plt.plot(self.dlist1,self.trace1)
55.         plt.xlabel("Distance from center")
56.         plt.ylabel("Radial Grav. Field")
57.         plt.tight_layout()
58.         plt.show()
59.         plt.savefig(img_filename)
60.
```

## Single-shell Model

```
1.  import math
2.  from grav_sim import grav_sim
3.  import matplotlib.pyplot as plt
4.
5.  class single_shell_model(grav_sim):
6.
7.      def __init__(self, r1, r2, M, Rx, Ry, Rz, fname):
8.          super().__init__(r1,r2,M,Rx,Ry,Rz)
9.          self.do_single_shell_system(r1, M, fname)
10.
11.     def do_single_shell_system(self,r1, M,fname):
12.         self.add_points_to_shell(r1)
13.         self.get_radial_gravitational_field()
14.         self.get_single_shell_ideal_traces(r1)
15.         self.plot_two_panel_single_shell_field(fname)
16.
17.     # Use fibonacci algorithm to distribute points on a sphere
18.     # from https://stackoverflow.com/questions/9600801/
19.     #         /evenly-distributing-n-points-on-a-sphere
20.
21.     def add_points_to_shell(self, shell_radius):
22.
23.         phi = math.pi * (math.sqrt(5.) - 1.)        # golden angle in radians
24.         N = self.n
25.         for i in range(N):
26.             y = shell_radius*(1 - (i / float(N - 1)) * 2)
27.             radius = math.sqrt(shell_radius**2 - y * y)    # radius at y
28.             theta = phi * i                               # golden angle incr.
29.
30.             x = math.cos(theta) * radius
31.             z = math.sin(theta) * radius
32.
33.             self.x.append(x)
34.             self.y.append(y)
35.             self.z.append(z)
36.
37.
38.     # Create the field where all the mass is at the center
39.     #
40.     def get_single_shell_ideal_traces(self,r1):
41.         self.dlist1,self.trace1 = self.get_ext_inv_r_squared_field(self.M, r1)
42.
43.
44.     def plot_two_panel_single_shell_field(self,img_filename):
45.         s = '[' + ','.join(str(x) for x in self.Obs ) +']'
46.         s = '    Sampling Vector: '+s
47.         self.make_parameter_string()
48.         model_name = "Single Shell Gravity Field: "+self.pstr + s
49.         mydpi=120
50.         fig = plt.figure(figsize=(1200/mydpi,600/mydpi),dpi=mydpi)
51.         plt.rcParams['axes.facecolor'] = 'white'
52.
53.         fig.suptitle(model_name)
54.         ax = fig.add_subplot(1,2,1,projection='3d')
55.         ax.scatter(self.x, self.y,self.z, color = 'blue',s=3)
56.         fig.add_subplot(1, 2, 2)
57.         plt.plot(self.dlist, self.g)
58.         plt.plot(self.dlist1,self.trace1)
59.         plt.xlabel("Distance from center")
60.         plt.ylabel("Radial Grav. Field")
61.         plt.subplots_adjust(wspace=.3)
62.         plt.show()
63.         plt.savefig(img_filename,dpi = mydpi)
64.
```

## Double-shell Model

```
1.  import math
2.  from grav_sim import grav_sim
3.  import matplotlib.pyplot as plt
4.
5.  class double_shell_model(grav_sim):
6.
7.      def __init__(self, r1, r2, M,Rx,Ry,Rz,fname):
8.          super().__init__(r1,r2,M,Rx,Ry,Rz)
9.
10.         self.do_double_shell_system(r1, r2, M, fname)
11.
12.
13.     def do_double_shell_system(self,r1, r2, M,fname):
14.         self.add_points_to_shell(r1, int(self.n/2))
15.         self.add_points_to_shell(r2, int(self.n/2))
16.         self.get_radial_gravitational_field()
17.         self.get_double_shell_ideal_traces(r1, r2)
18.         self.plot_two_panel_double_shell_fields(fname)
19.
20.
21.     # Use fibonacci algorithm to distribute points on a sphere
22.     # from https://stackoverflow.com/questions/9600801
23.     # /evenly-distributing-n-points-on-a-sphere
24.     def add_points_to_shell(self, shell_radius, np):
25.
26.         phi = math.pi * (math.sqrt(5.) - 1.)          # golden angle in radians
27.
28.
29.         for i in range(np):
30.             y = shell_radius*(1 - (i / float(np - 1)) * 2)
31.             radius = math.sqrt(shell_radius**2 - y * y)    # radius at y
32.             theta = phi * i                          # golden angle increment
33.
34.             x = math.cos(theta) * radius
35.             z = math.sin(theta) * radius
36.
37.             self.x.append(x)
38.             self.y.append(y)
39.             self.z.append(z)
40.
41.
42.     # Create the field where all the mass is at the center
43.     #
44.     def get_double_shell_ideal_traces(self,r1, r2):
45.         m = self.M/2
46.         self.dlist1,self.trace1 = self.get_ext_inv_r_squared_field(m, r1)
47.         self.dlist2,self.trace2 = self.get_ext_inv_r_squared_field(2*m, r2)
48.
49.
50.     def plot_two_panel_double_shell_fields(self,img_filename):
51.
52.         s = '[' + ','.join(str(x) for x in self.Obs ) +']'
53.         s = '    Sampling Vector: '+s
54.         self.make_parameter_string()
55.         model_name = "Double Shell Gravity Field: "+self.pstr +s
56.         mydpi=120
57.         fig = plt.figure(figsize=(1200/mydpi,600/mydpi),dpi=mydpi)
58.         plt.rcParams['axes.facecolor'] = 'white'
59.
60.         fig.suptitle(model_name)
61.         ax = fig.add_subplot(1,2,1,projection='3d')
62.         ax.scatter(self.x, self.y,self.z, color = 'blue',s=3)
63.         fig.add_subplot(1, 2, 2)
64.         plt.plot(self.dlist, self.g)
65.         plt.plot(self.dlist1,self.trace1)
66.         plt.xlabel("Distance from center")
67.         plt.ylabel("Radial Grav. Field")
68.         plt.show()
69.         plt.savefig(img_filename,dpi = mydpi)
70.
```

## Sphere

```
1.  import math
2.  import random
3.  from grav_sim import grav_sim
4.  import matplotlib.pyplot as plt
5.
6.  class sphere_model(grav_sim):
7.
8.      def __init__(self, r1, r2, M, Rx, Ry, Rz, fname):
9.          super().__init__(r1,r2,M,Rx,Ry,Rz)
10.         self.do_spherical_system(r1, M, fname)
11.
12.
13.     def do_spherical_system(self,r1, M, fname):
14.         self.add_points_to_sphere(r1,M)
15.         self.get_radial_gravitational_field()
16.         self.get_sphere_ideal_traces(r1)
17.         self.plot_two_panel_sphere_field(fname)
18.
19.
20.     def add_points_to_sphere(self, shell_radius,N):
21.
22.         r = shell_radius
23.
24.         count = 0
25.
26.         while count < N:
27.             x = r*(random.random()*2 - 1)
28.             y = r*(random.random()*2 - 1)
29.             z = r*(random.random()*2 - 1)
30.
31.             if (x*x + y*y + z*z) < r*r:
32.                 self.x.append(x)
33.                 self.y.append(y)
34.                 self.z.append(z)
35.                 count = count + 1
36.
37.
38.     def get_radial_gravitational_field_solids(self):
39.         o = self.Obs
40.         d_Obs = math.sqrt(o[0]**2 + o[1]**2 + o[2]**2)
41.         onx   = o[0]/d_Obs
42.         ony   = o[1]/d_Obs
43.         onz   = o[2]/d_Obs                       # radial unit vector
44.         res = d_Obs/self.Nsteps
45.         self.dlist = [x*res +res/10 for x in range(0,self.Nsteps)]
46.         m1 = self.M/self.n
47.         for d in self.dlist:
48.             gx = 0
49.             for n in range(0,self.n):
50.
51.                 dx       = d * onx              # sampling pos is [dx, dy, dz]
52.                 dy       = d * ony
53.                 dz       = d * onz
54.
55.                 sx       = dx - self.x[n]
56.                 sy       = dy - self.y[n]
57.                 sz       = dz - self.z[n]
58.                 s2       = sx**2 + sy**2 + sz**2
59.                 s        = math.sqrt(s2)
60.
61.                 if s > 2*res :
62.                     sn       = [sx/s, sy/s, sz/s]        # unit vector to Obs
63.
64.                     cos_theta = sn[0]*onx + sn[1]*ony + sn[2]*onz
65.                     grav      = m1/s2               # mass = m1 = M/n
66.                     gravr     = grav*cos_theta      # force radial component
67.                     gx        = gx + gravr
68.             self.g.append(gx)
69.
70.     # Create the field where all the mass is at the center
71.     #
72.     def get_sphere_ideal_traces(self,r1):
73.         self.dlist1,self.trace1 = self.get_ext_inv_r_squared_field(self.M, r1)
74.
```

```
75.    def plot_two_panel_sphere_field(self,img_filename):
76.        self.make_parameter_string()
77.        s = '[' + ','.join(str(x) for x in self.Obs ) +']'
78.        s = '    Sampling Vector: '+s
79.        model_name = "Ball Gravity Field: "+self.pstr +s
80.        mydpi=120
81.        fig = plt.figure(figsize=(1200/mydpi,600/mydpi),dpi=mydpi)
82.        plt.rcParams['axes.facecolor'] = 'white'
83.        fig.suptitle(model_name)
84.        ax = fig.add_subplot(1,2,1,projection='3d')
85.        ax.scatter(self.x, self.y,self.z, color = 'blue',s=3)
86.        fig.add_subplot(1, 2, 2)
87.        plt.plot(self.dlist, self.g)
88.        plt.plot(self.dlist1,self.trace1)
89.        plt.xlabel("Distance from center")
90.
91.        plt.ylabel("Radial Grav. Field")
92.        plt.show()
93.        plt.savefig(img_filename)
94.
```

## Ring Model

```
1.  import math
2.  from grav_sim import grav_sim
3.  import matplotlib.pyplot as plt
4.
5.  class ring_model(grav_sim):
6.      def __init__(self, r1, r2, M,Rx,Ry,Rz,fname):
7.          super().__init__(r1,r2,M,Rx,Ry,Rz)
8.          self.do_ring_system(r1, M,fname)                    #r2 unused
9.
10.     def do_ring_system(self,r,n, fname):
11.         self.radius = r
12.         self.M = n
13.         self.add_points_to_ring(r, n)
14.         self.get_radial_gravitational_field()
15.         self.get_ring_ideal_trace()
16.         self.plot_two_panel_ring_fields(fname)
17.
18.     def add_points_to_ring(self,r, N):
19.         dphi_deg = 360./self.M
20.         for i in range(0,N):
21.             phi_deg = i*dphi_deg + .1
22.             phi_rad = math.radians(phi_deg)
23.             self.x.append(r*math.cos(phi_rad))
24.             self.y.append(r*math.sin(phi_rad))
25.             self.z.append(0)
26.
27.     # Create the field assuming all the mass is at the center
28.
29.     def get_ring_ideal_trace(self):
30.         self.dlist1,self.trace1 = \
31.             self.get_ext_inv_r_squared_field(self.M, self.radius)
32.
33.     def plot_two_panel_ring_fields(self, img_filename):
34.         s = '[' + ','.join(str(x) for x in self.Obs ) +']'
35.         s = '    Sampling Vector: '+s
36.         self.make_parameter_string()
37.         model_name = "Gravity Field for Ring: " + self.pstr +s
38.
39.         mydpi=120
40.         fig = plt.figure(figsize=(1200/mydpi,600/mydpi),dpi=mydpi)
41.         plt.rcParams['axes.facecolor'] = 'white'
42.
43.         fig.suptitle(model_name)
44.         ax = fig.add_subplot(1,2,1,projection='3d')
45.         ax.scatter(self.x, self.y,self.z, color = 'blue',s=3)
46.         fig.add_subplot(1, 2, 2)
47.         plt.plot(self.dlist, self.g)
48.         plt.plot(self.dlist1,self.trace1)
49.         plt.xlabel("Distance from center")
50.         plt.ylabel("Radial Grav. Field")
51.
52.         plt.show()
53.         plt.savefig(img_filename)
54.
```

## SUMMARY

In this chapter, we modeled various particle spatial distributions to study their gravitational effects. Using them we could verify properties known from calculus and test models against ideal behaviors. But there are many we didn't attempt such as cubes, cylinders, and lines of particles, which would be easy extensions. Most importantly, we found critical models involving shells and spheres matched results expected from theory, which is not only useful, but supportive of insight.

Most of the models could be assembled into more complex ones – in fact, as we will see in the next chapter, we can model a spiral galaxy's structure by combining spherical and disk distributions. While we used models where sampling was done along axes of symmetry, which would usually be amenable to algebraic analysis, measurements along non-symmetric axes could be easily done, even though otherwise, they would probably be much more difficult, without easy analytic solutions. And of course, with simple modification, by allowing particles to have an electrical charge, many of the models could be used for studying charge distributions also.

## CHAPTER 5

# Spiral Galaxies and Dark Matter

WHEN ASTRONOMERS STUDY SPIRAL galaxies, one thing they measure is their rotation curves which show how fast the stars are moving at various distances out from the center. For example, on his website (https://w.astro.berkeley.edu/~mwhite/darkmatter/rotcurve.html), Professor Martin White includes a figure from the study by Begeman (1989) showing the rotation curve for galaxy NGC3198 (see Figure 5.1).

Astronomers quickly noticed an odd thing about such galaxy curves – they often flatten – which means the orbit speeds are not reducing with distance. This is interesting because for systems orbiting a central mass, such as the planets in our Solar System, the further out you go, the slower the planets move (see Figure 5.2). In fact, by not tapering off, the galaxy rotation curves suggest the outer stars are orbiting too fast and the galaxies should be flying apart!

Where things get really interesting is that when we model a spiral galaxy as having a core and a disk, this is not sufficient to explain the observations, there needs to be a third invisible component surrounding the whole galaxy which we call 'dark matter.'

We will now adapt the tools developed when studying mass distributions to explore galaxy rotation curves, by trying galaxy models with spherical cores and a disk, and then seeing if the resulting rotation curves can be adjusted as needed using a halo of matter mimicking dark matter, to give results similar to the observed rotation curves.

DOI: 10.1201/9781003600046-5

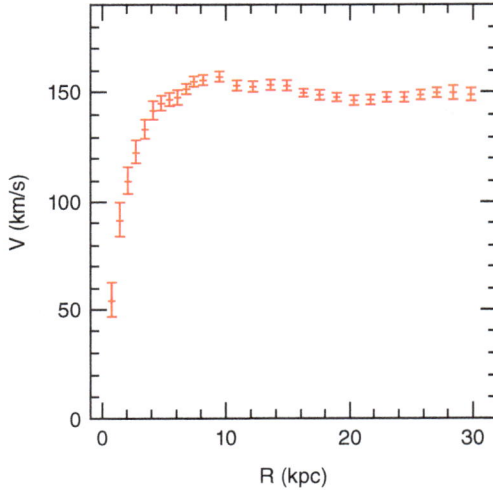

FIGURE 5.1    The rotation curve for galaxy NGC3198 shows how fast the stars are moving (orbiting the center) at various distances out from the center.

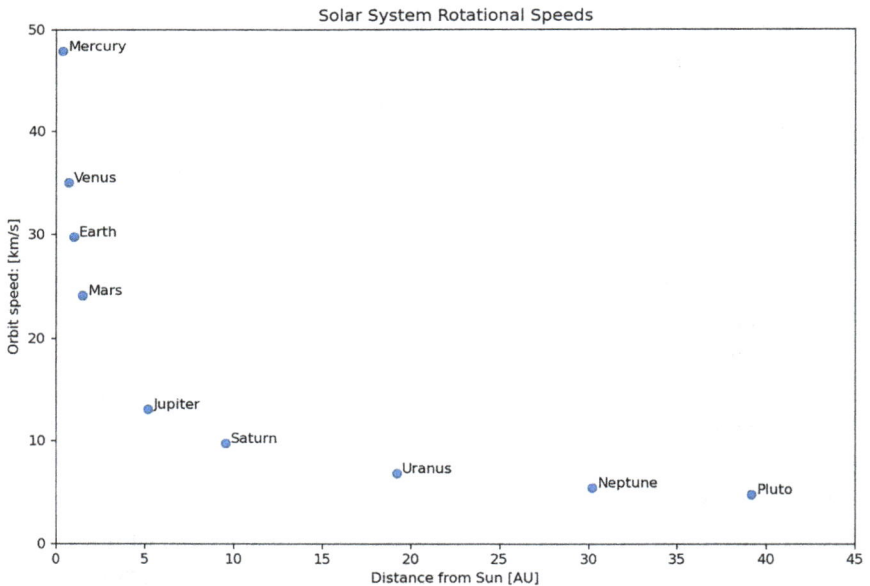

FIGURE 5.2    The rotation curve of the Solar System shows that the inner planets rotate around the Sun with faster velocities than the outer planets. Credit: NASA/SSU/Aurore Simonnet.

Because we are considering one particular class of galaxy models (core, disk, and halo) and because we will be considering velocity and not the gravitational field, we will create a new class and calculate orbital velocity instead of gravitational force, with the following elements:

1. There will be a new controlling class, class_spiral_galaxy_v.py instead of grav_sim.py.

2. The class will be self-contained and not use a parent class for simplicity.

3. The mass will be specified separately for the core, disk, and halo.

4. We will estimate the velocities along the x-axis for simplicity.

5. The velocity at a distance r from the center will be based on the normal rule for circular orbits:

$$v = \sqrt{\frac{GM}{R}} \qquad (5.1)$$

Equation 5.1 is usually derived by balancing centripetal and gravitational forces, and so M represents the mass at the center of the orbit. We are dealing with mass distributions and will rewrite the equation to reflect this:

$$v = \sqrt{\frac{GM}{R}} = \sqrt{\frac{RGM}{R^2}} = \sqrt{RF} \qquad (5.2)$$

In this form, we see the velocity is the square root of R times the gravitational field (force per unit mass). This means the effect of the mass distribution M creates a resulting gravitational field F at a distance R, and this is what's being balanced against the centripetal force to create the orbit. So, instead of using M to directly calculate v using Equation 5.1, we will, just as we did in the previous chapter's models, use M to calculate F, and from that we calculate v.

Our goals then are to achieve the following: create spiral galaxy mass distributions and to calculate their rotation curves. The mass distributions will be based on a spherical distribution at the center representing the galaxy core; a disk of material to represent the galaxy's disk; and a halo of material surrounding the core and disk. The core and disk will represent what we normally see, and the halo will be the extra mass needed to account for typical observed rotation curves. Our halo will be a shell with an inner and outer radius because we expect matter on the interior to have condensed into the core and disk, so by having an inner and an outer radius, we can explore the effects of different halo sizes and thicknesses.

Our code is contained in two files: **class_spiral_galaxy_v.py** and **run_models.py. class_spiral_galaxy_v.py** contains the formal definition for the class we created – **spiral_galaxy_model_v()**, and its details are discussed below. There you will find the complete code and detailed notes on how the code functions, such as how the rotational speeds were actually calculated (implemented), the galaxy models built, and how the graphics output was done using matplotlib plot libraries.

## RUNNING THE GALAXY MODELS

A run of a model is done by instantiating the class, with 4 size, 3 mass, and 3 position parameters. This gives us full control over all critical aspects of our model. Specifically, a model is run by calling the class method **spiral_galaxy_model_v()** in a file (we call **run_models.py**), which simply imports the class and instantiates a model. For example,

$$sv = spiral\_galaxy\_model\_v\left(2, 8, 30, 40, 200, 400, 3000, 50, 0, 0\right)$$

Sets R1–R4 as 2, 8, 30, and 40; M1, M2 and M3 as 200, 400, and 3000, respectively; and sets the end of the sampling radial on the x-axis (50, 0, 0).

When running a simulation, through invoking instantiations of the **spiral_galaxy_model_v** class, the following happens:

- Parameters R1–R4 are used to set the sizes of the different galaxy components: the core radius, the outer radius of the disk, and the inner and outer radii of the halo.

- Parameters M1, M2, and M3 set the masses (number of points) in each of the components.

- Parameters Rx, Ry, and Rz set the end point of the sampling radial. We use about 50 steps along the radial for the sampling points.

- Points – (x,y,z) coordinates – are generated for each of the components – there are M1 of them for the core, etc. so the model has a total of M1+M2+M3 points.

- For every sampling point on the sampling radial, the gravitational field there is calculated from all points in the galaxy and from that the rotation speed found using Equation 5.2. However, to avoid numerical instabilities, points very close to the sampling point are ignored.

- Ideal traces, where the enclosed (interior) mass is assumed to be at the center, are calculated for the core, the core and disk, and the whole system. In each case, the largest scale is used, R1, R2, and R4. The corresponding enclosed masses would be M1, M1+M2, and M1+M2+M3, respectively.

- A two-panel plot is created with a 3-D image of the galaxy on the left, and on the right, estimated and ideal traces for the rotation speeds displayed.

Multiple models (instantiations) can be specified in the **run_models.py** file, perhaps grouped to emphasize parameter combinations of interest.

Note: Every instantiation causes a new browser page to appear with that model's results.

## TESTING MODELS

In this section, we show the results for models where different combinations of core/disk/halo masses and sizes are tried, and the rotation curves calculated and displayed. At first, no halo will be set, so we can model the expected velocity fall off and then we add in halo mass to see if we get rotation curve flattening, and this of course is the fundamental argument justifying the existence of a halo, referred to as consisting of Dark Matter because it has a significant gravitational effect but was undetected prior to rotation curve studies.

Let's first look at two scenarios where we ignore the halo and create a disk equal to, and three times the core mass, as shown in Figures 5.3 and 5.4. For our Milky Way galaxy, the disk is about three times the core mass, and its radius is about 15 times larger. The left panels show the core and disk, color coded blue and green. The panels on the right show the estimated rotation speeds (red) and the theoretical from using only the core (blue dots) and their combined mass (gold dots).

In the equal mass scenario, the combined mass is twice that of the core, so the corresponding rotation speeds are root-two (1.41) times greater. When the disk is three times the core, the combined mass is four times

Spiral Galaxy Radial Velocity: (1,15,10,20,  200,200,0,  50,0,0)

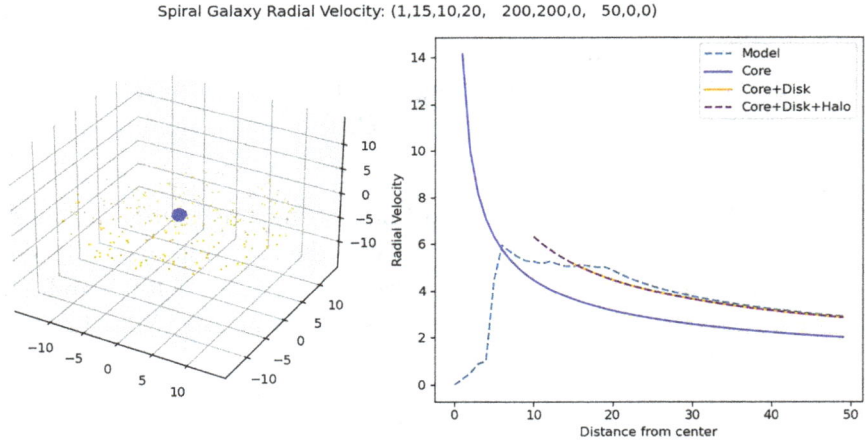

FIGURE 5.3   Rotation speeds when the disk equals the core mass. The red curve is computed from the mass distribution. Blue dots show the theoretical for the core and gold dots for the disk and core. Since rotation speed scales as the square root of the enclosed mass, outside the disk, the combined disk and core results are 1.41 (square root of two) that of the core only results.

Spiral Galaxy Radial Velocity: (1,15,10,20,  200,600,0,  50,0,0)

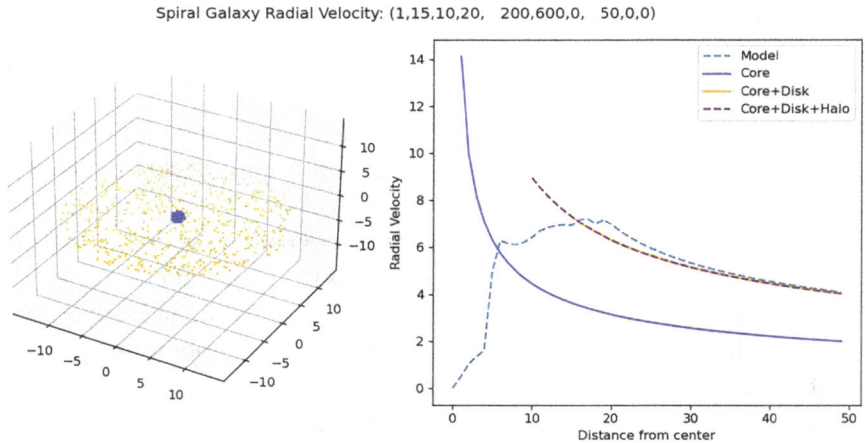

FIGURE 5.4   Increasing the disk mass to three times the core increases overall rotation speeds out to the disk edge, when the expected fall off occurs. Since the total enclosed mass outside the disk is four times that of the core, the speeds are twice that of just the core mass.

that for the core, and rotation speed scales as a factor of two, so, the exterior rotation speeds (gold dots) are generally twice that of the core's (blue dots).

Figures 5.3 and 5.4 also show that outside the disk, the whole system behaves as if all the mass is effectively at the center, since it so closely

Spiral Galaxy Radial Velocity: (1,25,10,20,  200,600,0,  50,0,0)

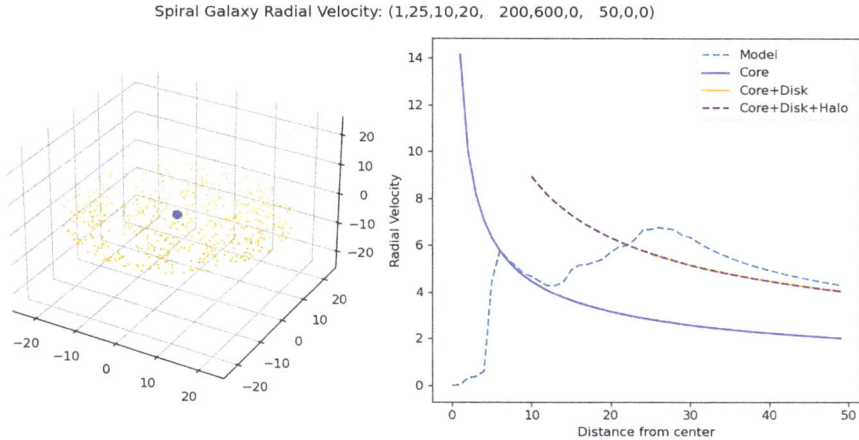

FIGURE 5.5   Spreading the disk mass over a larger disk of size 25 does flatten our rotation curve so this might work for some galaxies but would be inconsistent with our galaxy's disk size.

follows the ideal traces – consistent with our experiments in the previous chapter. Note, the **core+disk+halo** curve mostly overlaps that for the **core+disk** in these plots.

These scenarios also show that while we can add more mass to the disk, the disk cannot produce a flattened rotation curve to the outside. More mass must be added outside the disk.

While it wouldn't be consistent with our galaxy's disk, a galaxy with a disk 25 times the core's size, would produce a relatively flat rotation curve as shown in Figure 5.5, where the disk was increased from 15 to 25 times the size of the core.

So, if we want to flatten the rotation curve beyond the disk, we need to add additional mass over and beyond what we see in the form of the core and disk – called 'dark matter.' Figure 5.6 shows a scenario where additional mass is added overlapping the disk (R3 = 10, R4 = 40) and flattens the rotation curve, giving a more uniform overall appearance.

By being able to model core, disk, and halos with different masses and sizes, these models give much room to match observed rotation curves, but it must be remembered that models should be based on observed characteristics. In all cases, once we reach the edge of a system, the rotation curve must take on the ideal form, and if it doesn't, then that indicates there is still more mass unaccounted for.

For completeness, the above charts were produced by the run_models. py file shown in Figure 5.7.

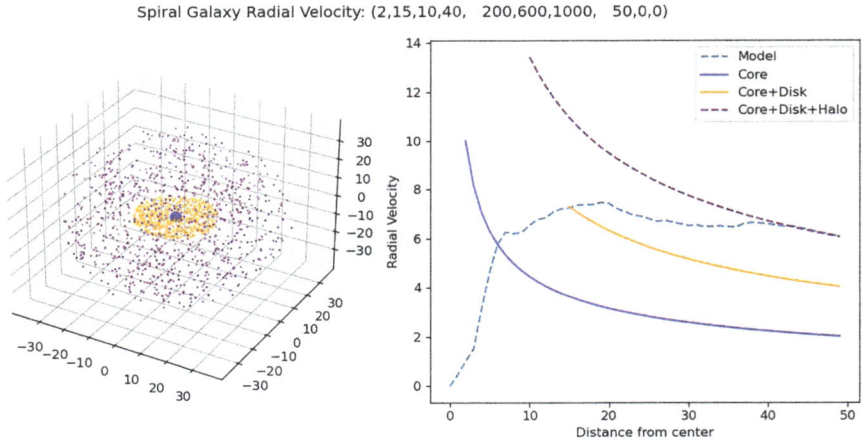

FIGURE 5.6   With a thicker dark matter halo extending from 10 to 40, overlapping the disk significantly, the rotation curve is smoother and flatter.

```
1. from class_spiral_galaxy_v import spiral_galaxy_model_v
2.
3.
4. sv = spiral_galaxy_model_v(1,15,10,20, 200, 200, 0, 50,0,0,'./Fig 5.3.jpg')
5. sv = spiral_galaxy_model_v(1,15,10,20, 200, 600, 0, 50,0,0,'./Fig 5.4.jpg')
6.
7. sv = spiral_galaxy_model_v(1,25,10,20, 200, 600, 0, 50,0,0,'./Fig 5.5.jpg')
8. sv = spiral_galaxy_model_v(2,15,10,40, 200, 600, 1000, 50,0,0,'./Fig 5.6.jpg')
9.
```

FIGURE 5.7   The **run_models.py** file used to generate this chapter's plots.

There are many possible variations on our models appropriate as projects for undergraduate students. For example, the following could be investigated:

- Modify the code so sampling along non-radial paths could be done.

- Investigate transverse fields along the sampling radial.

- Try combinations of geometries: multiple rings or parallel disks above and below the x-y plane.

- Explore the gravitational/electrical potential. (Hint: divide by r instead of $r^2$ when calculating particle effects.)

- Modify the force calculation sign. In the gravity models, the positive sign indicated an attractive force. However, many will prefer positive to suggest a force pushing outward.

- In our galaxy models, no attempt was made to provide any kind of density gradient or structure to the mass distributions, so the models could be improved by adding points based on density gradient formulae.

- There is evidence for both a thin and a thick disk in our Galaxy that might be included.

- There have also been papers discussing an inner halo inside the disk radius, so a second halo could be added to model this effect.

- Some galaxies have ring structures – these could be modeled by modifying the disk creation routine to only select points between selected radii.

## CLASS spiral_galaxy_model_v PROGRAMMING NOTES

We use the same Python and Plot libraries as before. The simulation could be invoked by simply instantiating the class and letting the __**init**__() function call the model. Because our models require multiple parameters to be specified, these will be passed through from the instantiation and used by the __**init**__() to set them internally for the class.

Unlike our previous models (e.g., the double-shell system), the mass of each component is specified uniquely (as **M1**, **M2**, and **M3**, and not as the total mass), and we use four size parameters (**R1–R4**) to set the core and disk radii, and the inner and outer halo shell radii. As before, **Rx**, **Ry**, and **Rz** are the coordinates of the sampling radial limit, and we set the number of samples along the radial as **self.Nsteps** (lines 19–20).

For this class, there are three ideal traces – one for each of the components – based on an enclosed mass ideal. This means that for the disk, we assume all the mass is inside **R2** and consists of **M1+M2**, and that for the halo we assume all the mass is inside **R4** and includes the total mass **M1+M2+M3**.

The **do_spiral_galaxy_radial_v**() is the method used by the class to process a model (see line 23). It builds the core; adds the disk; builds the halo as a thick shell; computes the ideal and estimated orbit velocities; and creates the output plots. It is invoked when a model is instantiated as part of the __**init**__() function.

To build the three parts of the spiral galaxy mass distribution, we use two class methods. To create the spherical core and the halo, we use **add_ points_to_shell**() where an inner and outer radius is set, along with the number of points to use (i.e., the mass). For the spherical mass at the core, the inner radius is zero, but the halo can be constructed with the inner radius being non-zero. The disk is created using the **add_points_to_disk()** method with a radius and mass specifier.

(Note that the code could be modified to allow the disk to have an inner radius if annular structures were being investigated – structures such as ring galaxies, or in other contexts, planetary ring systems.)

The rotation curves are at the heart of our simulation, and these are calculated using the **get_rotation_curve()** (line 68) method, which calculates the orbital speeds by first calculating the radial gravitational field and then applying Equation 5.2 (line 98).

Because random sampling can result in getting too close to a point, as before, we exclude mass points too close to a sampling point to avoid them overly influencing the calculations. For samples very close to the origin, the gravitational field can be negative (outward in our convention), making the velocity calculation fail; as a work around, the absolute value is used in the velocity calculation at line 98 to handle the small number of points affected. Alternatively, this could be removed and the model rerun so a different set of points is generated and likely avoids the problem.

To avoid sampling points being overly influenced by nearby point masses, those within five sampling resolution increments are excluded (line 92).

Similar to our previous code for calculating the gravitational field, we now use the field to get the orbital speed. The general idea here is that at each mass point located at vector **r_vec**, a vector (**s_vec**) to the sampling point at **d_vec** is constructed (lines 86–88). Unit vectors of each (**rn**, and **sn**) are constructed for convenience, since the dot product between **rn** and **sn** gives the cosine of the angle between them – needed to get the field contribution along the sampling radial (line 94). The rotation speed can then be calculated from the net gravitation field and the radial distance.

We learned in a previous chapter how all mass distributions such as shells, rings, and disks have exterior gravitational fields that trend $M/r^2$ where the mass is effectively at the center. This approximation can be used to estimate rotational velocities providing us with a powerful cross-check; an analytic solution against which our results can be compared. Function **get_ext_inv_r_squared_rotation()** (line 102) calculates the radial velocities for the specified mass beyond the specified radius. Theoretical ideal curves can then be generated for the core, disk, and halo, where the mass used is the mass enclosed by each structure, which is why for the halo, we use the outer radius **self.r4** instead of the inner radius (**self.r3**).

The two-panel plot is created by the **plot_two_panel_spiral_galaxy_rotation()** which plots the mass distribution on the left (line 163), and on the right side, the modeled velocity and the theoretical curves (line 167).

To prevent auto-scaling from distorting the galaxy mass plot, function **set_limits()** ensures all axes limits are set to the maximum of all coordinates (line 137).

Also, **make_parameter_string** creates a string listing model parameters to be used in the chart caption.

```
 1.  import math
 2.  import random
 3.  import matplotlib.pyplot as plt
 4.
 5.  class spiral_galaxy_model_v(object):          # a class to generate rotation curves
 6.
 7.      def __init__(self, r1, r2, r3, r4, M1, M2, M3,Rx,Ry,Rz,fname):
 8.          self.r1          = r1            # core
 9.          self.r2          = r2            # disk
10.          self.r3          = r3            # halo inner
11.          self.r4          = r4            # halo outter
12.          self.M1          = M1
13.          self.M2          = M2
14.          self.M3          = M3
15.
16.          self.M           = M1+M2+M3
17.          self.n           = self.M
18.
19.          self.Obs         = [Rx,Ry,Rz]    # sampling radial end-point
20.          self.Nsteps      = 50            # radial points to sample
21.          self.do_spiral_galaxy_radial_v(fname)
22.
23.      def do_spiral_galaxy_radial_v(self,fname):
24.          self.x = []; self.y = []; self.z = []
25.          self.add_points_to_shell(0,self.r1, self.M1)       # core
26.          self.add_points_to_disk(self.r2, self.M2)          # disk
27.          self.add_points_to_shell(self.r3,self.r4, self.M3) # halo
28.          self.get_rotation_curve()
29.          self.get_spiral_galaxy_ideal_rotations()
30.          print("plotting...")
31.          self.plot_two_panel_spiral_galaxy_rotation(fname)
32.
33.      def add_points_to_shell(self, rmin, rmax, n):
34.          rsqr_min = rmin*rmin
35.          rsqr_max = rmax*rmax
36.          N = n
37.          count = 0
38.          while count < N:
39.              x = rmax*(random.random()*2 - 1)
40.              y = rmax*(random.random()*2 - 1)
41.              z = rmax*(random.random()*2 - 1)
42.              rsqr = x*x + y*y + z*z
43.
44.              if (rsqr < rsqr_max) and (rsqr > rsqr_min):
45.                  self.x.append(x)
46.                  self.y.append(y)
47.                  self.z.append(z)
48.                  count = count + 1
49.
50.      def add_points_to_disk(self, disk_radius,M):
51.
52.          r = disk_radius
53.          N = M
54.          count = 0
```

```
55.
56.          while count < N:
57.              x = r*(random.random()*2 - 1)
58.              y = r*(random.random()*2 - 1)
59.              z = 0
60.
61.              if (x*x + y*y ) < r*r:
62.                  self.x.append(x)
63.                  self.y.append(y)
64.                  self.z.append(z)
65.
66.                  count = count + 1
67.
68.      def get_rotation_curve(self):
69.          o = self.Obs
70.          d_Obs     = math.sqrt(o[0]**2 + o[1]**2 + o[2]**2)
71.          onx       = o[0]/d_Obs
72.          ony       = o[1]/d_Obs
73.          onz       = o[2]/d_Obs                  # radial unit vector
74.          res       = d_Obs/self.Nsteps
75.          N         = len(self.x)
76.          self.dlist = [x*res  for x in range(0,self.Nsteps)]
77.          m1        = self.M/self.n
78.          self.v    = []
79.          for d in self.dlist:
80.              gx = 0
81.              for n in range(0,N):
82.                  dx      = d * onx            # sampling pos is [dx, dy, dz]
83.                  dy      = d * ony
84.                  dz      = d * onz
85.
86.                  sx      = dx - self.x[n]
87.                  sy      = dy - self.y[n]
88.                  sz      = dz - self.z[n]
89.                  s2      = sx**2 + sy**2 + sz**2
90.                  s       = math.sqrt(s2)
91.
92.                  if s > 5*res :
93.                      sn        = [sx/s, sy/s, sz/s]     # unit vector to Obs
94.                      cos_theta = sn[0]*onx + sn[1]*ony + sn[2]*onz
95.                      grav      = m1/s2            # mass = m1 = M/n
96.                      gravr     = grav*cos_theta       # force radial component
97.                      gx        = gx + gravr
98.              rvel = math.sqrt(abs(d*gx))
99.
100.             self.v.append(rvel)
101.
102.     def get_ext_inv_r_squared_rotation (self, total_mass,radius ):
103.
104.         o = self.Obs
105.         d_Obs = math.sqrt(o[0]**2 + o[1]**2 + o[2]**2)
106.
107.         res = d_Obs/self.Nsteps
108.         self.dlist = [d*res for d in range(0,self.Nsteps)]
109.
110.         dl = []
111.         tr = []
112.
113.         for d in self.dlist:
114.             if d >= radius:
115.                 tr.append(math.sqrt(total_mass/d))
116.                 dl.append(d)
117.         return dl, tr
118.
119.
120.     # Create the velocity fields assuming all interior the mass is at the center
121.     #
122.     def get_spiral_galaxy_ideal_rotations(self):
123.         M12 = self.M1 + self.M2
124.         M123 = self.M1 + self.M2 + self.M3
125.         self.dlist1,self.vtrace1 = self.get_ext_inv_r_squared_rotation(self.M1, self.r1)
126.         self.dlist2,self.vtrace2 = self.get_ext_inv_r_squared_rotation(M12, self.r2)
```

```
127.              self.dlist3,self.vtrace3 = self.get_ext_inv_r_squared_rotation (M123, self.r3)
128.
129.
130.      def make_parameter_string(self):   # make a string like [r = 1,0, M = 500]'
131.
132.              rlist = str(self.r1)+','+str(self.r2)+','+str(self.r3)+','+str(self.r4)
133.              mlist = str(self.M1)+','+str(self.M2)+','+str(self.M3)
134.              Rlist = str(self.Obs[0])+','+str(self.Obs[1])+','+str(self.Obs[2])
135.              self.pstr = '(' + rlist + ',   ' + mlist + ',   '+Rlist +')'
136.
137.      def set_limits(self,ax):           # used to draw corners to force axis scales
138.              dmax = max(self.x + self.y+ self.z)
139.              print('dmax is: ', dmax)
140.
141.              ax.set_xlim([-dmax, dmax])
142.              ax.set_ylim([-dmax, dmax])
143.              ax.set_zlim([-dmax, dmax])
144.
145.
146.      def plot_two_panel_spiral_galaxy_rotation(self,img_name):
147.              c1 = ['blue']*self.M1
148.              c2 = ['orange']*self.M2
149.              c3 = ['purple']*self.M3
150.              clist = c1 + c2 + c3
151.              print("Making rotation plot")
152.              self.make_parameter_string()
153.              model_name = "Spiral Galaxy Rotation Curve: "+self.pstr
154.              print(model_name)
155.
156.              self.make_parameter_string()
157.              model_name = "Spiral Galaxy Radial Velocity: "+self.pstr
158.              mydpi=120
159.              fig = plt.figure(figsize=(1200/mydpi,600/mydpi),dpi=mydpi)
160.              plt.rcParams['axes.facecolor'] = 'white'
161.
162.              fig.suptitle(model_name)
163.              ax = fig.add_subplot(1,2,1,projection='3d')
164.              self.set_limits(ax)
165.              ax.scatter(self.x, self.y,self.z, color = clist,s=0.5)
166.
167.              fig.add_subplot(1, 2, 2)
168.              plt.plot(self.dlist, self.v, linestyle='dashed',label="Model")
169.              plt.plot(self.dlist1,self.vtrace1, color='blue',label="Core")
170.              plt.plot(self.dlist2,self.vtrace2, color='orange',label="Core+Disk")
171.              plt.plot(self.dlist3,self.vtrace3, linestyle='dashed', \
172.                       color='purple',label="Core+Disk+Halo")
173.              plt.xlabel("Distance from center")
174.              plt.ylabel("Radial Velocity")
175.              plt.legend()
176.              plt.tight_layout()
177.              plt.show()
178.              plt.savefig(img_name, dpi=mydpi)
179.
180. if __name__ == '__main__':
181.
182.      sv = spiral_galaxy_model_v(2,8,20,40, 200, 400, 0,    50,0,0,'./Fig A.jpg')
183.      sv = spiral_galaxy_model_v(2,8,20,40, 200, 400, 3000, 50,0,0,'./Fig B.jpg')
184.
```

## SUMMARY

In this chapter, we developed a collection of tools and classes to explore the gravitational fields created by various geometrical mass distributions. Similar models, based on these, could be derived to explore the electric fields associated with charge distributions. The models were important for two reasons. First, they show how to 'brute force' solutions, when analytical solutions relying on integral calculus are too difficult; and second, they provide an opportunity to cross-check results from calculus, for example, to independently verify that indeed gravitational forces disappear inside a spherical shell.

We also explored how with a simple core and disk structure, we cannot account for the observed rotation curves seen in many spiral galaxies, and we found we could account for the discrepancy by adding additional matter in the form of larger halos of dark matter, far outside the observed light/star distribution.

## CHAPTER 6

# Sampling a Distribution

I N THIS CHAPTER, WE will explore some techniques that can be used to create samples based on probability distributions that are not necessarily uniform. For example, detailed studies of globular clusters, such as M13 shown in Figure 6.1, show that their density profiles are not typical, and not Gaussian, and require the researcher develop new, highly customized models. We provide an overview of essential concepts will be provided, and standard models developed to show how they function and to

FIGURE 6.1 M13, a famous globular cluster in the constellation Hercules. Such clusters can have hundreds of thousands of stars with spatial density distributions that are not normally encountered in textbooks, and the usual tools for creating Gaussian or exponential distributions are inadequate.

DOI: 10.1201/9781003600046-6

help the reader develop their own versions to learn from. For us, a model will refer to a system's Probability Distribution Function (PDF). Once we have our models, we can then calculate the Cumulative Distribution Functions (CDFs), the inverse CDFs, and demonstrate how to use the inverse CDFs to generate a sample consistent with the initial PDF.

## PDFs AND CDFs

A PDF is a function p(x) that says what the likelihood of something happening is, over an interval. If x is a random variable, then p(x)$\Delta$x is the probability of finding x between x and x+$\Delta$x. For a uniform distribution, p(x) = k a constant – it doesn't depend on x, so all x outcomes are equally likely. But what if we want more small values than large ones? A uniform distribution, by itself, won't work.

Suppose we wanted to try a linear model where p(x) behaved like 1–x, so when x was zero, p() would be large, but when x was 1, p() would be zero. A rule like p(x) = 2 (1–x) would work. The factor of 2 is needed so the area under the p(x) line adds up to unity – the total probability must always add up to one. In the language of calculus, we would say the integral of p(x)dx is one, that is, $\int p(x)dx = 1$, which is the same as saying the area under the p(x) curve must be 1. We should also note, for example, that this particular rule allows for p(x) to equal 2 when x is zero; the probability density can exceed unity, but the total probability cannot!

Most software systems like Python provide libraries with built-in basic probability density distributions, such as for uniform or Gaussian (bell curve) distributions. But what if you want something other than these? Like one for our globular cluster above?

One solution is to first get the probability density's CDF (Cumulative Distribution Function), invert the CDF, and then use a random number generator with the inverted CDF to select x-values. We will now explain this in detail.

A CDF is a function where CDF(x) is the area under p(x) from infinity up to x. In words, CDF(x) is the probability that the random variable is less than x. For example, if CDF(2) equals 0.4, then the probability that x is less than 2 is 0.4 or 40%. For a uniform p(x) = 1, where x is in [0, 1], the CDF(x) is simply x, since the integral of 1 is x.

It might be helpful to compare probability with mass density. Metals have densities of about 10g/cc. It makes no sense to point at a spot on a car body and ask how much mass is there. You must combine volume with density to figure out the mass, so you might additionally specify something like 'in that 5cc part of the car'. The answer would be 10g/cc x 5cc = 50g. And similarly with the PDF; p(x) by itself doesn't tell us what the probability is, but p(x)Δx does.

The CDF takes us from the world of probability densities and differentials into the world of probabilities. For example, what's the probability of x falling between A and B where B > A? Answer: CDF(B) – CDF(A). So, the CDF is a function that essentially calculates percentiles.

Many books on statistics will provide a proof of why sampling the inverse CDF randomly can produce a sample consistent with the CDF; a proof that usually only takes a few lines and is mathematically quite straightforward.

Let's say X is a random variable such that

$$X = G(U) \tag{6.1}$$

In other words, we can generate a collection of X values by using a collection of uniform values U and the function G(). Then, for any X < x, it must be true that

$$
\begin{aligned}
& X && < x \\
\Rightarrow \quad & G(U) && < x \\
\Rightarrow \quad & U && < G^{-1}(x)
\end{aligned}
$$

Now it follows that

$$Pr(X < x) = Pr\left(U < G^{-1}(x)\right)$$

But the left side is just F(x) – the CDF – and for a uniform distribution, since $Pr(A < B) = B$, the right side is $G^{-1}(x)$, therefore

$$F(x) = G^{-1}(x) \tag{6.2}$$

Therefore, the function G(U) needed to create a sample of X is $F^{-1}()$, the inverse CDF.

So, when generating X from the inverse CDF of a random variable using $F^{-1}(U)$ in Eq. 6.1, X has a CDF given by F (Eq. 6.2), which is what we want. Since uniform random number generators are relatively well understood and readily available, they can be used to provide the U values used by the $F^{-1}$ functions to create the desired X distribution samples.

It is hoped that by seeing plots of the PDF, CDF, inverse CDF, and the corresponding histograms of data generated from the inverse CDF, the reader will become more comfortable with this fussy process, especially when the process is applied to a diverse collection of distributions like the uniform, linear, quadratic, and Gaussian, and seen to be successful on each.

We can use the CDF to generate sample sets consistent with the associated PDF, since unlike the PDF, there is a unique x for every CDF(x). (With a PDF, such as a Gaussian distribution, low likelihood events can occur at either end of the distribution, so a PDF value is not necessarily unique.) This means we can work backwards. For example, if we use a random number generator to create a set of y values $\{y_1, y_2, y_3...\}$ in the range [0,1], each $y_i$ can be considered a percentile where CDF($x_i$) = $y_i$ for some unique $x_i$, and so the set $\{x_1, x_2, x_3...\}$ is a sample consistent with the PDF used to create the CDF. Finding the $x_i$, where CDF($x_i$) is $y_i$, requires inverting the CDF; if y = 3x, then to find the x for a given y we would solve x = y/3, and swap the x and y labels to get the inverse functional form: y = x/3. Therefore, if we have a CDF that we can invert to get CDF$^{-1}$, then we can create a set of sample events $\{x_i\}$, from a set of random values between 0 and 1, using $x_i$ = CDF$^{-1}$($y_i$). In words, a random sampling of the inverse CDF results in a set of {xi}distributed according to the system's PDF.

Some of this can be confusing when first encountered, in part because the definition of the CDF involves an integral. But the integral is simply a mechanism to get the area under the PDF; if the PDF was flat, we could bypass the integral and get the area using simple geometry. Ultimately, the CDF associates an event, $x_i$, with its percentile, $y_i$. Perhaps another example will help. Let V(x) be the area of a cube of side x, then V(X) = $x^3$. What's the inverse function? How do we get the side when the volume is v? Answer: S(v) = $v^{1/3}$. S() and V() are inverses of each other.

What about the clever discussion about the CDF(a) being the probability x is less than or equal to 'a'? Not a problem. Let's replace V(x) = $x^3$ with H(x) = $x^3$ and say H(x) is the largest volume that can fit in a cube of size x. With this change, we have simply added a new interpretation to what the equation form means, but the underlying mechanics have not changed at all, so don't let a nuanced interpretation confuse things. Ultimately, the CDF is just a function that maps events to percentiles. As for the integral aspect of CDF, we could say our V(x) function was the integral of the volume contained within the cube; it might sound scarier and more sophisticated, but it's still $x^3$.

To elaborate on the above concepts, we will now consider a suite of distribution possibilities to see what their PDFs, CDFs, CDF$^{-1}$s look like and

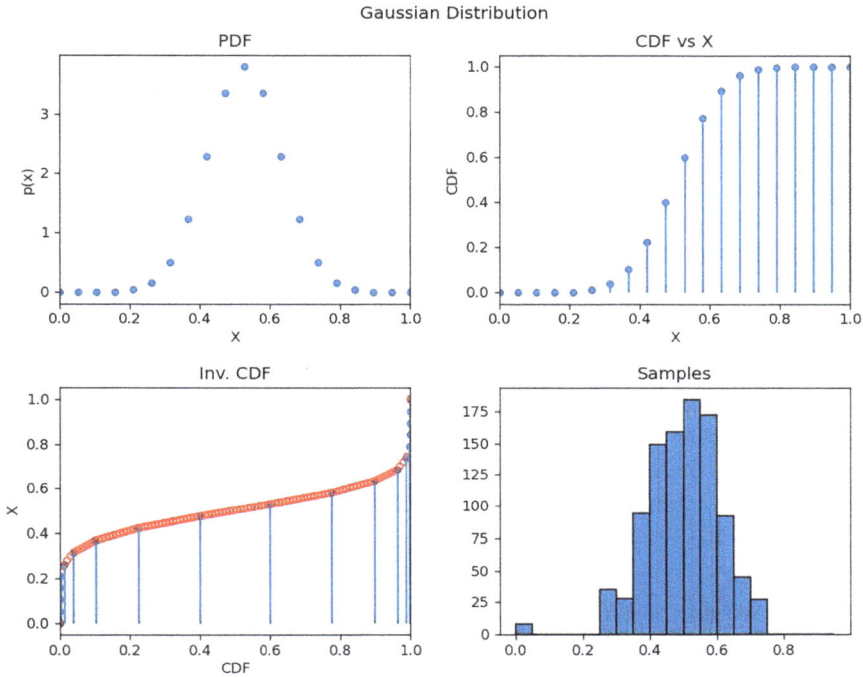

FIGURE 6.2    A chart produced by running **model_dist** in Gaussian mode.

see if a sample set drawn from sampling a CDF$^{-1}$ is consistent with the underlying PDF, and the code to do this is in class **model_dist** which produces a graphical output for a Gaussian PDF as shown in Figure 6.2.

The plots in Figure 6.2 show the Gaussian PDF in the top left corner, and its CDF in the top right. The inverted CDF (i.e., CDF$^{-1}$) is shown in the bottom left, and a histogram of samples generated from sampling CDF$^{-1}$ in the lower right. Note how the inverted CDF plot has unevenly spaced X-axis values. This is remedied by resampling, and the result shown as the red overlay. It is also worth noting with the PDF plot, that for each x, there is a unique p(x), but this not true for the inverse, most p(x) values have two possible x values, which makes it difficult to use p(x) directly for creating a sample of x.

Class **model_dist**'s functionality breaks down into the following areas:

1. Configuring PDFs of interest

2. Calculating the corresponding CDFs

3. Calculating the inverse CDFs

4. Generating a sample dataset using the inverse CDF

5. Displaying the results

These subtasks will be described below and were implemented in a hopefully very straightforward fashion to make them as transparent as possible for the reader. The random variables are all in in the interval [0, 1], and the sampling resolution was fixed at 20 samples. Calculations were mostly based on an approximation where the area under a curve is simply the height times the interval even though an analytic solution would be more accurate; of course, increased precision could be achieved by increasing the number of sampling points.

To emphasize some of the underlying issues here, consider a uniform distribution where x is in [0, 1]. Since the integral of the area must be 1, p(x) is 1. How should we model this very simple distribution in our computer code? Let's try using equally spaced sampling points such as {0, 0.2, 0.4, 0.6, 0.8, 1}; dividing the domain into fifths produces six sampling points. Two are at the boundaries (0, and 1.0), From calculus, the area would be the sum of p(x)*0.2 = 0.2*(1+1+1+1+1) = 0.2 * 5) = 1. Note we omitted the last boundary point at x = 1.0. To include that would have us including an area element outside the [0, 1] domain. What if we had a set of random values, which we strongly suspected were uniformly distributed? We might have to use a convention to ignore the area element associated with the largest. In any case, the larger the number of points, the smaller the consequences of including or omitting the largest point. Going from the perfectly analytic to the sampled digital can introduce granularity in results – usually handled by increasing the number of samples and or sometimes by adding corrections to your algorithms.

There are a variety of models available: Uniform ('u'), Linear Increasing ('li'), Linear Decreasing ('ld'), Gaussian ('g'), and Quadratic ('q').

A model is run by instantiating the class, with a desired model type, for example, for a Gaussian model, a command like: **sa = model_dist('g')** will run the model and produce the four-panel plot display shown in Figure 6.2.

## CLASS model_dist PROGRAMMING NOTES

Class **model_dist** (see Figure 6.3) is best understood by viewing it as a collection code blocks (function groups) that perform the following roles: Initialization and model selection; **PDF, CDF, CDF⁻¹** creation; data manipulation – resampling and sampling; and plot/subplot creation.

```
 1. import matplotlib.pyplot as plt
 2. import numpy as np
 3. from numpy import random as npr
 4. import math
 5.
 6. class model_dist:
 7.     def __init__(self,mode,fname):
 8.         self.N  = 20                        # number of points
 9.         self.NB = 5*self.N                   # sample histogram bins
10.         self.make_pdf(mode)
11.         self.make_cdf()
12.         self.make_inv_cdf()
13.         self.get_samples()
14.         self.make_four_panel_chart(fname)
15.
16. ################################ make PDF, CDF, and inv_CDF ##########
17.
18.     def make_pdf(self,mode):
19.         if mode == 'g':
20.             x, pdf = self.make_gaussian_pdf()
21.             title='Gaussian Distribution'
22.         elif mode == 'u':
23.             x, pdf = self.make_uniform_pdf()
24.             title='Uniform Distribution'
25.         elif mode == 'li':
26.             x, pdf = self.make_linear_i_pdf()
27.             title='Increasing Linear Distribution'
28.         elif mode == 'ld':
29.             x, pdf = self.make_linear_d_pdf()
30.             title = 'Decreasing Linear Distribution'
31.         elif mode == 'q':
32.             x, pdf = self.make_quadratic_pdf()
33.             title = 'Quadratic Distribution'
34.         self.main_title = title
35.         self.x    = x
36.         self.pdf  = pdf
37.
38.     def make_cdf(self):
39.         cdf = np.zeros(self.N)
40.         dX = 1/(self.N-1)
41.         cdf[0] = self.pdf[0]*dX
42.         for i in range(1,self.N):
43.             val = self.pdf[i]*dX
44.             cdf[i] = cdf[i-1]+val
45.         self.cdf = cdf
46.
47.     def make_inv_cdf(self):
48.         x0     = self.cdf
49.         y0     = self.x
50.         self.resample(x0,y0)
51.
52. ############################### make specific models ##############
53.
54.     def make_uniform_pdf(self):        # uniform
55.         pdf = np.zeros(self.N)
56.         X   = np.zeros(self.N)
57.         dX  = 1/(self.N-1)
58.         for i in range(0,self.N):
59.             pdf[i] = 1
60.             X[i] = i*dX
61.         return X, pdf
62.
63.     def make_linear_i_pdf(self):        #linear increasing
64.         pdf = np.zeros(self.N)
65.         dX   = 1/(self.N-1)
66.         X    = np.zeros(self.N)
67.         for i in range(0,self.N):
68.             pdf[i] = i/(self.N-1)
69.             X[i] = i*dX
70.         s = 0.5*pdf[self.N-1]
71.         return X, pdf/s
72.
73.     def make_linear_d_pdf(self):        # linear decreasing
74.         pdf = np.zeros(self.N)
```

FIGURE 6.3    Class **model_dist** used to demonstrate inverse sampling for various distribution types.

*(Continued)*

```
75.         dX   = 1/(self.N-1)
76.         X    = np.zeros(self.N)
77.         for i in range(0,self.N):
78.             pdf[i] = 1 - i/(self.N-1)
79.             X[i] = i*dX
80.         s = 0.5*pdf[0]
81.         return X, pdf/s
82.
83.     def make_quadratic_pdf(self):        # quadratic
84.         data = np.zeros(self.N)
85.         dX   = 1/(self.N-1)
86.         X    = np.zeros(self.N)
87.         for i in range(0,self.N):
88.             data[i] = ((i/(self.N-1))**2)
89.             X[i] = i*dX
90.         s = 1/3
91.         return  X, data/s
92.
93.     def make_gaussian_pdf(self):        # gaussian
94.         sig = .1
95.         a   = 0.5
96.         data = np.zeros(self.N)
97.         X = np.zeros(self.N)
98.         dX = 1/(self.N-1)
99.         for i in range(0,self.N):
100.            arg = -0.5*((i/self.N - a)/sig)**2
101.            data[i] = math.exp(arg)
102.            X[i] = i*dX
103.         s = sum(data)*dX
104.         return X, data/s
105.
106. ###################################### generate subplots ##############
107.
108.    def plot_pdf(self):
109.        plt.xlim([0, 1])
110.        plt.scatter(self.x, self.pdf, s=20)
111.        plt.title('PDF')
112.        plt.xlabel('X')
113.        plt.ylabel('p(x)')
114.
115.    def plot_cdf(self):
116.        bins = np.linspace(0,1,self.N)
117.        plt.xlim([0, 1])
118.        plt.scatter(bins, self.cdf, s=20)
119.        plt.title('CDF')
120.        plt.xlabel('X')
121.        plt.ylabel('Pr(X < x)')
122.
123.    def plot_cdf_stem(self):
124.        plt.xlim([0, 1])
125.        plt.stem(self.x, self.cdf, markerfmt=' ',basefmt=' ')
126.        plt.title('CDF vs X')
127.        #self.axes[r,c].set_yticks(self.cdf)
128.        plt.xlabel('X')
129.        plt.ylabel('CDF')
130.        plt.scatter(self.x, self.cdf, s=20)
131.
132.    def plot_inv_cdf_stem(self):
133.        plt.xlim([0, 1])
134.        plt.stem(self.cdf, self.x, markerfmt=' ',basefmt=' ')
135.        plt.title('Inv. CDF')
136.        plt.xlabel('CDF')
137.        plt.ylabel('X')
138.        plt.scatter(self.cdf, self.x, s=20)
139.
140.    def plot_resampled(self,a,b):
141.        plt.scatter(a, b, s=20, facecolors='none', edgecolors='r')
142.
143.    def plot_sample_hist(self):
144.        a = self.xs
145.        bins = np.arange(0,1,.05)
146.        plt.title('Samples')
147.        plt.hist(a,bins = bins,edgecolor='black')
148.
149. ############################### resample and generate samples ##############
```

FIGURE 6.3 (CONTINUED)  Class **model_dist** used to demonstrate inverse sampling for various distribution types.

*(Continued)*

```
150.
151.     def get_samples(self):  # sampling inverse cdf
152.         N2 = 1000
153.         yr = self.yr
154.         values = npr.randint(0,self.NB, N2)
155.         x1 = np.zeros(N2)
156.         for i in range(0,N2-1):
157.             x1[i] = yr[values[i]]
158.         self.xs = x1
159.
160.     def resample(self,x0, y0):
161.         NR = self.NB
162.
163.         range0 = max(x0) - min(x0)
164.         delta = range0/(NR-1)
165.
166.         yr = np.zeros(NR)
167.         xr = np.zeros(NR)
168.         xr[0] = x0[0]
169.         yr[0] = y0[0]
170.         yr[NR-1] = y0[self.N-1]
171.         xr[NR-1] = x0[self.N-1]
172.
173.         idx = 0
174.         for i in range(1, NR -1):
175.             xr[i] = delta*i + xr[0]
176.             j = idx
177.             while xr[i] > x0[j]:
178.                 j = j + 1
179.                 if j > self.N:
180.                     break
181.             idx = j-1
182.             dx = x0[idx+1] - x0[idx]
183.             f = (xr[i] - x0[idx] )/ dx
184.             #yr[i] = 0.5*(y0[idx+1] + y0[idx])        # simple average option
185.             yr[i] = y0[idx] + f* (y0[idx+1] - y0[idx])
186.
187.         self.xr = xr
188.         self.yr = yr
189.
190. ################################### make 4-panel chart #################
191.
192.     def make_four_panel_chart(self,img_fname):
193.         mydpi=120
194.
195.
196.         fig=plt.figure(figsize=(1200/mydpi,1000/mydpi),dpi=mydpi)
197.         fig.subplots_adjust(wspace=.1, hspace=.5)
198.         fig.suptitle(self.main_title)
199.
200.         plt.subplot(2, 2, 1)
201.         self.plot_pdf()
202.         plt.subplot(2, 2, 2)
203.         self.plot_cdf_stem()
204.         plt.subplot(2, 2, 3)
205.         self.plot_inv_cdf_stem()
206.         self.plot_resampled(self.xr,self.yr)
207.         plt.subplot(2, 2, 4)
208.         self.plot_sample_hist()
209.         plt.subplots_adjust(left=0.1,
210.                             bottom=0.2,
211.                             right=0.9,
212.                             top=0.9,
213.                             wspace=0.4,
214.                             hspace=0.4)
215.         plt.show()
216.         plt.savefig(img_fname,dpi=mydpi)
217.
218. if __name__ == '__main__':
219.
220.     sa = model_dist('li','./Fig Lin. Incr. Model.jpg')
221.     sa = model_dist('ld','./Fig Lin. decr. Model.jpg')
222.     sa = model_dist('q','./Fig Quadratic Model.jpg')
223.     sa = model_dist('u','./Fig Uniform Model.jpg')
224.     sa = model_dist('g','./Fig Gaussian Model.jpg')
225.
```

FIGURE 6.3 (CONTINUED)   Class **model_dist** used to demonstrate inverse sampling for various distribution types.

For a selected model, the mode (model) is set (see line 7) and the initialization builds the corresponding **PDF**, **CDF**, and **CDF**$^{-1}$ arrays with the **make_pdf**, **make_cdf**, and **make_inv_cdf** functions. After this, it creates a sample set by sampling the inverse CDF array (**get_samples**) and generates the chart output (**make_four_panel_chart**).

Note that once a model is specified and its PDF is created, all other functions are generic in the sense they work for all models/PDFs. This means plots can be readily produced for empirically based **PDF**s and other non-standard distributions.

In running a model, a variety of arrays are produced with self-explanatory names: **self.x**, **self.pdf** and **self.cdf**; inverse mapping is simply done by reversing the **self.x** and **self.cdf** arrays when necessary. As discussed later, **self.xr** and **self.yr** are also created based on resampling of the **self.x** and **self.cdf** arrays because of uneven spacing/resolution introducing unwanted granularity.

A critical aspect of the **make_inv_cdf** function is the order in how it passes **self.cdf** and **self.x** to the **self.resample()** function, that is, it passes in the arrays as (cdf, x) not (x, cdf), and this reversal is how the CDF inversion is achieved.

We will now look at the results for the different distributions to see how the inverse sampling processes work.

## Uniform Distribution

With a uniform distribution, (see Figure 6.4), all outcomes (events) are equally likely, so a random variable with values in [0, 1] would have a uniform distribution if its PDF was of the form p(x) = K, where K is a constant. In this case, K = 1, so the area under p(x) is also 1. The histogram of the generated samples is shown, which is flat – consistent with the PDF.

PDF values were calculated for evenly spaced x values between [0, 1], so p(x) is always one. CDF(x) is the area under p(x) between 0 and x, and hence is proportional to x, resulting in a straight line. Note the maximum value CDF(1) = 1. The histogram of the sample values used 20 bins for the 1000 points used in sampling the random variable which are distributed fairly evenly.

While not needed for the Uniform PDF, all models used resampling to handle uneven spacing of CDF values. The resampling function (**self.resample**) works by mapping the unevenly separated x values onto a larger number of evenly spaced ones defined over the same interval and linearly estimating the new y-values. For example, if the original data had two

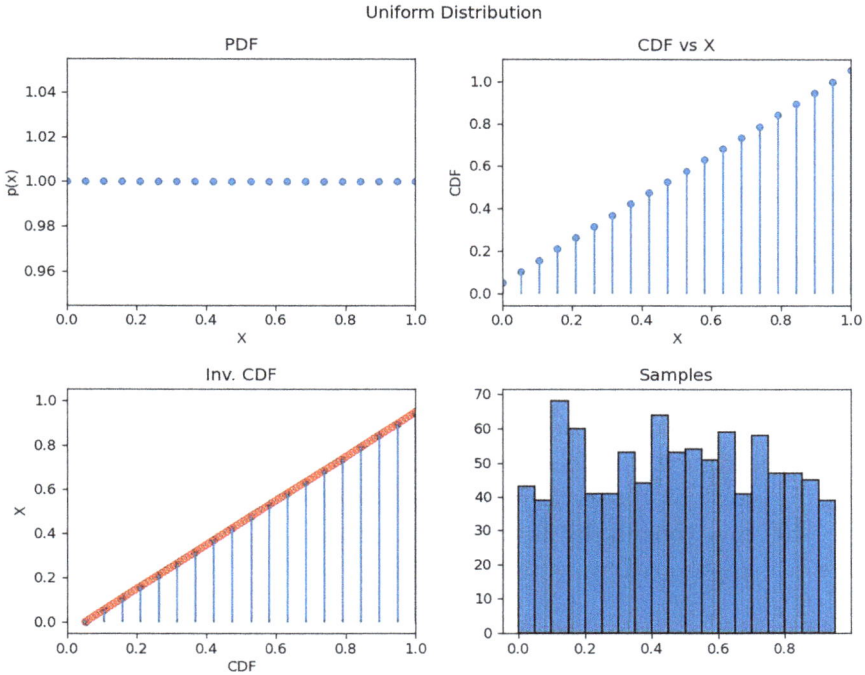

**FIGURE 6.4** Results for a uniform distribution. The PDF function p(x) is constant (top left); the CDF and the inverse CDF are linear (top right and bottom left), and the histogram of the generated samples is flat (bottom right). The red dots are for the resampled inverse CDF.

consecutive points (x1, y1) and (x2, y2), and then an oversampling point between them (x0,y0) would have a y-value, y0 = y1 + f*(y2−y1), where f was (x0 − x1)/(x2−x1). In other words, whatever fraction of the way x0 was in the x interval, so also was the y0 value in the y interval. The results of the resampling are shown as the red dots in the lower left chart panels and are seen to fill in the gaps in the original CDF values.

Now that we have our inverse CDF in the form of **xr**[] and yr[], it is easy to generate the samples consistent with the initial PDF using the **self.get_ samples** function (see Figure 6.5).

```
1.   def get_samples(self):  # sampling inverse cdf
2.       N2 = 1000
3.       yr = self.yr
4.       values = npr.randint(0,self.NB, N2)
5.       x1 = np.zeros(N2)
6.       for i in range(0,N2-1):
7.           x1[i] = yr[values[i]]
8.       self.xs = x1
9.
```

**FIGURE 6.5** Generating a sample from the CDF⁻¹ data.

To understand how the **get_samples** function works, remember, in our models, the **x**, **pdf**, and **cdf** arrays each have 20 elements. The resampled CDF$^{-1}$ arrays **xr** and **yr**, each have 1000 elements. To randomly pick 1000 numbers consistent with **yr**, an array (**values[]**) of length 1000 is created, in which each element is randomly assigned a number between 0 and 99 (see line 4). Then an array of 1000 samples **xs[]** is created from **xs[i]** = **yr[values[i]]**. The array **xs[]** is our desired sample set consistent with the model PDF.

## LINEAR MODELS

We now consider two linear models where the PDFs increase or decrease linearly – to perhaps be used when either a concentration or a dilution was needed near x = 0. The results of the models are shown in Figures 6.6 and 6.7. The pdfs look correct as do the samples generated. It is worth noting the **cdf()** functions in the bottom left panels are not evenly distributed – an effect emphasized by using a stem plot.

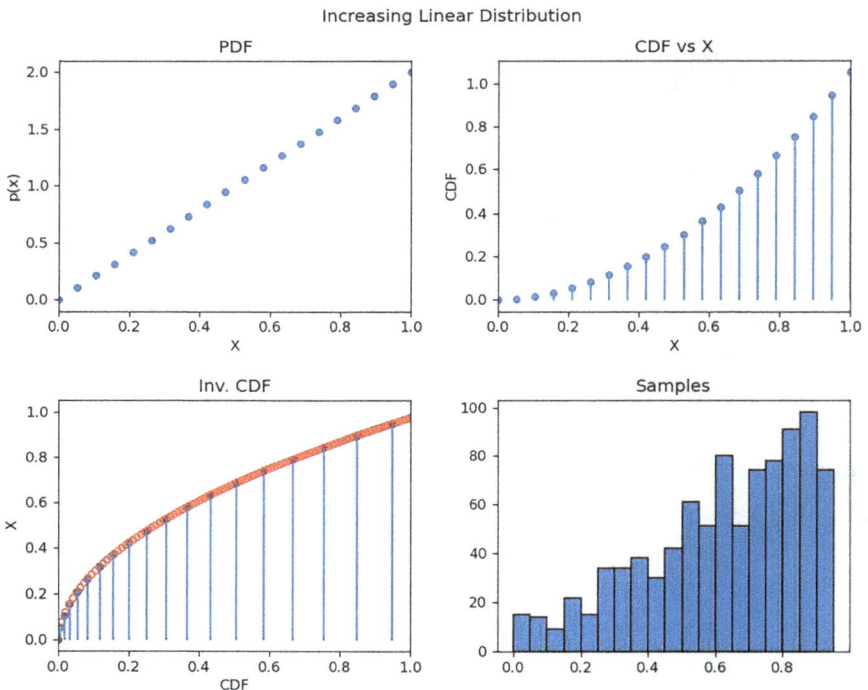

FIGURE 6.6    Results for the linearly increasing model.

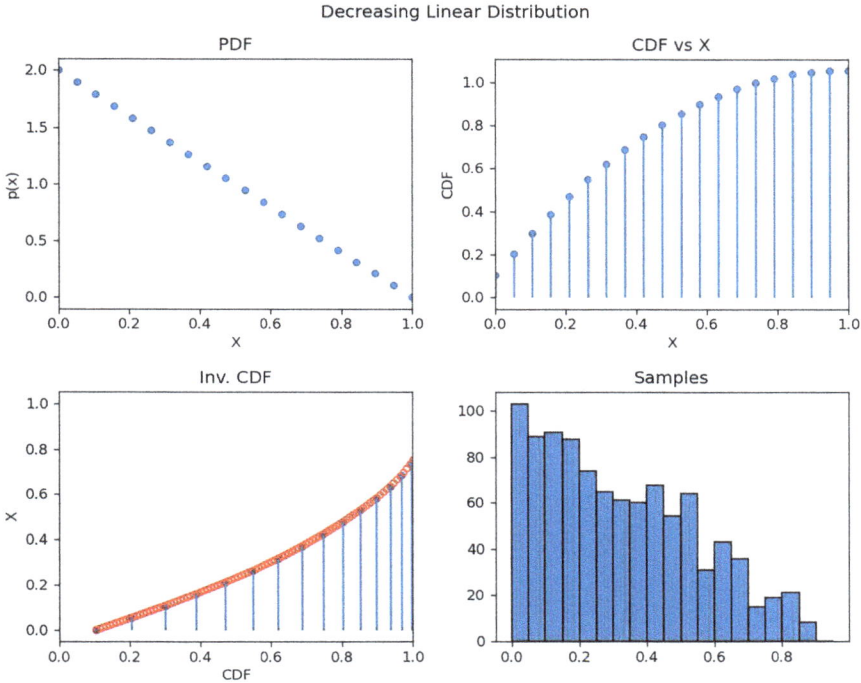

FIGURE 6.7    Results for the linearly decreasing model.

Analytically, the CDF for the linearly increasing pdf should scale as the integral of x (i.e., ½ x²) and for the linearly decreasing as the integral of (1–x) (i.e., x – ½ x²). In both cases, the CDF is not evenly spaced which is why we used resampling.

However, the red dots for the resampling look quite effective and match the raw inverse CDFs well, and the resampled inverse CDFs were used to create the samples.

## QUADRATIC AND GAUSSIAN PDFs

Because power laws and Gaussian/inverse Gaussian distributions are widely used, we will now consider their PDFs. For a quadratic power law, we will simply scale according to $x^2$, and for the Gaussian, as $\exp(-0.5)(x-a)^2/s^2$ where the center is at x=a=0.5, and the standard deviation is s=0.1. The results for the Gaussian model were already shown as Figure 6.2, but those for the quadratic are shown in Figure 6.8.

Both the quadratic and Gaussian results are as expected, with the latter being an extreme example of uneven spacing needing resampling.

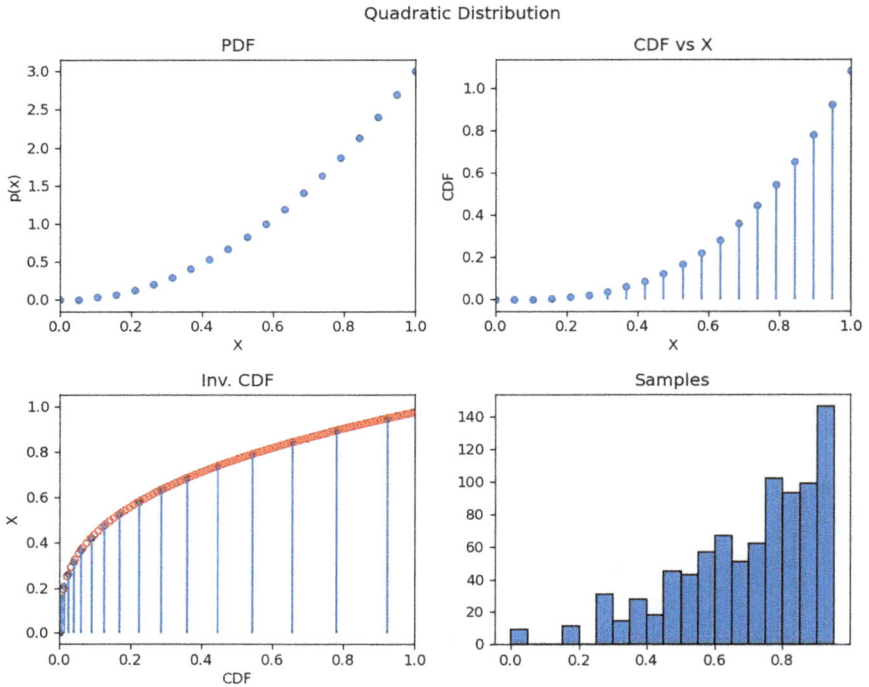

FIGURE 6.8   Results for the quadratic model.

## SUMMARY

In this chapter, we explored the problem of how to generate a sample from a distribution and illustrated the solution by applying it to a collection of standard forms (PDFs). The demonstration code could be modified and improved upon, but is functional, and most interestingly, the code could be used with non-standard PDFs, perhaps from lab experiments, which are not simple linear, quadratic, or Gaussian types.

The models showed how sampling an inverse CDF could produce a sample set consistent with the specified PDF, and the results were very reasonable and informative. One of the nice features of the code shown was that once a PDF was defined (i.e., **self.pdf**), all the other functions needed to create the CDFs and charts were immediately applicable. Also, by showing how to model uniform, linear, quadratic, and Gaussian distributions, not only do we illustrate how powerful the overall method was, but we provide a useful set of templates to build upon.

In the next chapter, we will tackle a problem taught in introductory college physics courses – projectile motion – but applied to very high velocity projectiles where aerodynamic drag changes the acceleration continuously and results in actual performance being considerably less than the theoretical.

## CHAPTER 7

# Projectiles – The German 88

A CLASSIC PROBLEM TAUGHT IN introductory college Physics courses is that of the projectile – motion in two dimensions under the influence of gravity. In this chapter, we will review the ideal solution taught in college Physics courses and apply it to a famous WWII artillery piece, the German 88. We will then modify our models to take drag into account and tune the model to match the gun's observed performance.

## PROJECTILE KINEMATICS WITH CONSTANT ACCELERATION

At the Earth's surface, the gravitational acceleration is g = −9.8 m/s/s and is vertical. Normally the projectile motion calculations assume there is no air resistance, and so there is no other acceleration/deceleration present. For a projectile, we have a launch angle θ and an initial velocity V. Because we choose cartesian coordinates (ones that are perpendicular to each other), and because we choose the y-axis to be vertical, both the x- and y- motions are independent of each other (other than at the start and stop) and gravity only impacts the y-motion.

The projectile problem simply applies the main equations from one-dimensional kinematics (the study of motion) to the two-dimensional case. In one-dimensions, we have the standard result:

$$x = v_i\, t + \tfrac{1}{2}\, a\, t^2 \qquad (7.1)$$

DOI: 10.1201/9781003600046-7

$$v = v_i + a\,t \tag{7.2}$$

$$v^2 = v_i^2 + 2\,a\,x \tag{7.3}$$

Here, the initial velocity is $v_i$, a is the acceleration, x is the distance travelled, and t is the elapsed time.

For a projectile, we assume there is a launch angle $\theta$, and an initial velocity V, then the initial component velocities are $v_x = V\,Cos(\theta)$ and $v_y = V\,Sin(\theta)$. In the x-direction, since a = 0, we simply have:

$$x = v_x\,t \tag{7.4}$$

and in the y-direction, we have:

$$y = v_y\,t + \tfrac{1}{2}\left(-g\right)t^2 \tag{7.5}$$

$$v(y) = v_y + \left(-g\right)t \tag{7.6}$$

$$v(y)^2 = v_y^2 + 2\left(-g\right)y \tag{7.7}$$

How high can the projectile go? At the high point v(y) = 0 m/s, and Eq. 7.7 yields:

$$H = \frac{v_y^2}{2g} \tag{7.8}$$

This makes sense: The faster the initial vertical velocity, the higher it should go, with gravity trying to reduce the effect. How long will the projective be airborne? The Time in Flight (T) is found from when the projectile is back on the (y = 0) and Equation 7.5 yields:

$$0 = t\left[v_y + \tfrac{1}{2}\left(-g\right)t\right] \tag{7.9}$$

In Eq. 7.9 we factored out one power of t for convenience, since in this form there are two ways to get zero on the left side: Either t is zero or the term in square brackets is zero. The first corresponds to the launch and the second to the striking the ground. Using the square bracket term, we can see that:

$$v_y + \tfrac{1}{2}(-g)\,T = 0, \tag{7.10}$$

Hence,

$$T = \frac{2v_y}{g} \tag{7.11}$$

So, the faster the vertical launch speed component, the longer the flight.

How far will the projectile travel (the range, R) before striking the ground?

$$
\begin{aligned}
R &= v_x T \\
&= V_x\, 2V_y / g \\
&= 2V^2 \cos(\theta)\sin(\theta)/g \\
&= (V^2/g)\sin(2\theta)
\end{aligned}
\tag{7.12}
$$

since from trigonometry, Sin(2A) = 2 Sin(A) Cos(A).

Equation 7.12 is reasonable: The range depends on the launch angle and launch speed. Note, since Sin(X) is a maximum when X is 90 degrees, the maximum range happens when $2\theta = 90$ or when $\theta = 45$ degrees.

Our equations for H, T, and R work well when we do not need to worry about friction or air-resistance, such as for low-velocity situations. However, when velocities are large, air resistance can be significant, and unlike gravity, varying, that is, changing with velocity. We will now explore this effect by analyzing a famous artillery piece from WWII, the German 88, first by comparing reported performance with our zero-resistance H, T, and R estimates, and then by seeing how we might build models to handle varying accelerations in both the x and y directions, to match observations.

## THE GERMAN 88

The German 88 (see Figure 7.1) was an 88-mm-bore artillery piece that fired very fast rounds, which meant they could generally strike targets with faster impacts and at further distances. Because it was such an effective weapon, it could be used in an anti-aircraft role as a flak gun or mounted on a tank. The flak version could also be used as an anti-tank

FIGURE 7.1  The German 8.8 cm Flak 36 gun; one of the most effective and versatile used in WWII.

weapon, killing enemy tanks at ranges where the enemy guns were ineffective. Because its shells were so fast, they could strike targets like tanks using a very flat trajectory making them easy to aim, and quickly – there was little need to lead the target. Also, the high muzzle velocity allowed them to strike at high flying aircraft.

While the performance might change depending on the type of shell used, some typical reported effective performance measures include: If V = 840m/s, H = 9900m, and R = 14860m, when θ is 90 and 45 degrees, respectively. Although there is some ambiguity as to what constitutes an effective range or ceiling in a military as compared to our kinematics' sense, these numbers will serve as useful references.

What do our equations for H, R, and T produce for the German 88? When θ = 45 degrees, the range R is 72000m and max height H = 18000m, with a time in flight T = 121s. When θ is 90 degrees, R is zero of course, H = 36000m, and T is 171s. Clearly, these estimates greatly exceed reported ones. The obvious explanation is that air resistance (drag) must be playing a critical role by adding a deceleration. So, if we wish to improve our estimates, we will need to take drag into account.

We know from aerodynamics that drag depends on speed and also on air density, and this presents an interesting challenge: How should we estimate drag since it will depend on both the projectile speed and altitude, and is changing continuously?

## KINEMATICS WITH AERODYNAMIC DRAG

Aerodynamic drag is a force that can be modelled by an equation of the form:

$$D = \tfrac{1}{2}\, C\, \rho\, A\, v^2 \tag{7.13}$$

where C is a constant, $\rho$ is air density, A is cross-sectional area, and v is speed. C can be estimated from aerodynamical/fluid mechanical calculations or measured in a wind-tunnel. In kinematics, we work with accelerations, not forces, and since acceleration is force divided by mass, we can re-write Eq. 7.13 to yield the acceleration from drag as:

$$a_D = \tfrac{1}{2}\left(C/m\right)\rho\, A\, v^2 \tag{7.14}$$

where m is the projectile mass. If we assume the projectile has a constant cross-sectional area, and a constant mass, and if we write the density as $\rho = f\,\rho_0$ where $\rho_0$ is sea-level density, we can simplify further:

$$a_D = \left[ \frac{1}{2}\left(C/m\right) A\, \rho_0 \right] f\, v^2 \tag{7.15}$$

hence,

$$a_D = K\, f\, v^2 \tag{7.16}$$

In this form, K combines all the physical properties of the shell (mass, cross-sectional area, drag coefficient) and sea-level density, and the acceleration (really a deceleration since it opposes motion) only depends on the speed, and f(), the density at a given height as a fraction of sea-level density. So, if we choose a density profile for the atmosphere by specifying f(), for any projectile, there will be a K that controls the deceleration. In other words, if we select K (perhaps through trial and error) to match one of the observations for a given launch angle, we should be able to reasonably model the behavior for all other launch angles.

Note that because f() will be smaller than one, and since velocity is large, $v^2$ can be very large – initially about $840^2$ for the German 88, K must be a small number, measured in parts per million.

Unlike Equations 7.4 through 7.8, we need to solve a system where the acceleration is changing in response to height and speed, so these equations won't work. However, if we break down the problem of modelling the trajectory into time-slices (intervals), we can reasonably assume the acceleration is constant during each interval, and we could use similar equations if a deceleration from drag is added to both the x and y motions. The strategy would be to calculate $\mathbf{a_D}$ at the beginning of an interval, based on the velocity and height, and use this to calculate the velocity, position, and height at the end of the interval; the ending values are then used as the starting values for the next interval, and the process is repeated for as many intervals needed to follow the trajectory path.

## INVESTIGATING THE GERMAN 88 GUN'S PERFORMANCE

The software solution developed to model projectiles with drag is a class called **projectile**. The software can plot charts and also text tables to the console. To apply it to the German 88 gun, the following steps were taken:

1. Verify the **K**=0 (drag-free) scenario produces results matching the theoretical for a given launch configuration (e.g., 840 m/s and angle 45 degrees).

2. Explore **R**, **H**, and **T** values for a wide range of **K** values to demonstrate how **K** affects the results.

3. Through trial and error, find the best **K** that matches our test scenario.

4. Using the best **K** value, explore performance measures (**R**, **H**, and **T**) for max range and max height.

5. Explore how the projectile velocity changes with time and distance.

To run the code, a command like:

$$p = \mathbf{projectile}()$$

is used to create an instance of the class, and then running

$$\mathbf{p.run\_K\_list}\left(840, 45, \left[0, .001\right]\right)$$

```
1. K=0.000000, V= 840, theta=45, R= 71989, H=17999, T=121
2. K=0.001000, V= 840, theta=45, R=  6087, H= 3266, T= 42
3.
```

FIGURE 7.2    Results displayed on the console for K = 0 and 0.001.

in the console will produce text output to the console as shown in Figure 7.2, where the **R, H,** and **T** values associated with **K** equal to 0, and .001, are shown:

As a test of the code, we will assume **K=0** for no drag, and the results should match the theoretical. First an instance of the class is created by p = projectile(), then **p.run_K_list(840,45,[0])** results in:

$$K = 0.000000, V = 840, theta = 45, R = 71989, H = 17999, T = 121$$

which agrees with our previous theoretical calculation.

Since the **K=0** results show very unreasonable ranges and heights for the German 88 gun, we need to try to figure out if there is a K value that will give results that match the observation. To study a range of **K** values, when θ is 45 degrees, we can run a script like that in Figure 7.3 where **K** is sampled at various powers of 10. The for-loop simply creates a list of **K** value strings for the program.

We know from observation that the max range is 14860 m, so the output is telling us that **K** must be less than 0.0001 since R is too small otherwise. By a process of trial and error, the closest match was found with K=0.00036. We can now compare the 45 degree max range and 90 degree max height scenarios for this K with the theoretical values from the Python console (see Figure 7.4):

The results are quite good. While the theoretical (no-drag, K=0) calculations were more than 400% wrong (e.g., ranges of 70km vs 15km), our models are within about 3% of the reported values.

```
 1. k_list=[0,.000001, .00001, .0001, .001]
 2.
 3. k_strs=[]
 4.
 5. for k in k_list:
 6.     k_strs.append(f"{k: >.6f}")
 7.
 8.
 9. p.run_K_list(840,45,k_list)
10. K=0.000000, V= 840, theta=45, R= 71989, H=17999, T=121
11. K=0.000001, V= 840, theta=45, R= 71430, H=17919, T=120
12. K=0.000010, V= 840, theta=45, R= 67051, H=17218, T=117
13. K=0.000100, V= 840, theta=45, R= 38878, H=12246, T= 96
14. K=0.001000, V= 840, theta=45, R=  6087, H= 3266, T= 42
15.
```

FIGURE 7.3    Running a script from the console to test different K values.

```
1. p.run_K_list(840,45,[.00036])
2. K=0.000360, V= 840, theta=45, R= 15468, H= 6643, T= 65
3.
4. p.run_K_list(840,90,[.00036])
5. K=0.000360, V= 840, theta=90, R=    0, H= 9733, T= 78
6.
```

FIGURE 7.4    Results for the 45 degree and 90 degree scenarios with K = 0.00036. Since we are exploring a specific K value, the K list has only one entry.

Now that we have a reasonable, functioning model, we can use the **projectile** class to explore the effects of drag in more detail. Using our code, a set of three plots was generated, showing **y** vs **x**, **v** vs **t**, and **v** vs **x**, with a **K** value list of [0, 0.000001, 0.00001, .0036, 0.001]. The results are shown in Figure 7.5. The results (top panel) show that as drag increases, the trajectory becomes less symmetric, the maximum height reduces and

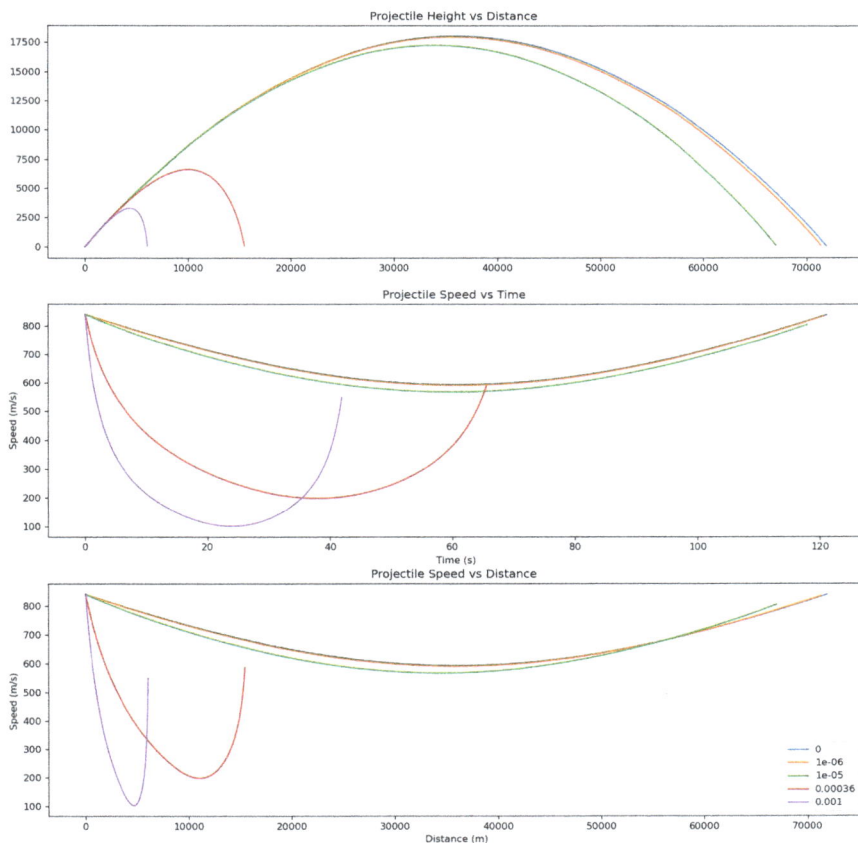

FIGURE 7.5    Model results for 45 degree launch angles. Height vs Distance (top), Speed vs Time (middle), and Speed vs Distance (bottom).

range decreases, and the descent from maximum height is steeper. The middle and bottom panels show how the speed changes throughout the flight. For the German 88 (**K**=.0036), the velocity starts at more than twice the speed of sound; becomes subsonic by 20s into its 65s flight; attains a minimum speed of about 200 m/s at the highest point; and then approaches almost twice the speed of sound as it falls under the influence of gravity.

We now have a reasonable model for projectile motion with drag present, and it's accurate to about 3%. In some ways, we have a surprisingly good result since we have seen the projectile for the German 88 experiences both supersonic and sub-sonic modes of flight, and having a single **K** value that seems to work for both modes is fascinating. But the **K** value we found through trial and error was a compromise trying to simultaneously match best range for $\theta$ = 45 degrees and max height for $\theta$ = 90 degrees. We could improve on this by allowing **K** to be a linear function of $\theta$, based on a K(45) = A, and K(90) = B where A and B are determined as before through matching observations using trial and error. Then, when updating the acceleration components, we would calculate K by:

$$K = \frac{B-A}{45}\theta + 2A - B \tag{7.17}$$

When $\theta$ is 45 and 90, K is A and B, respectively.

It would also be interesting to examine the case where the gun was used as an anti-tank weapon. What elevations (launch angles) are needed for various target distances? What were the times of flight? How far would a target move during the time in flight? This is left as an exercise for the reader.

Finally, our model can obviously be applied to guns other than the German 88. The **K** values will be different because of shells having different mass, shapes, and surface texture, and we would also assume because of them having different spin rates. Different projectiles could make for interesting studies depending on their launch speeds and intent – the infantry mortar would be an example of a subsonic projectile that might be very well modeled because of the simpler physics.

## CLASS PROJECTILE PROGRAMMING NOTES

To analyze and model projectile motion with high launch velocities, we need to include the effects of drag-induced decelerations. Our code has a basic OOD structure and is shown in Figure 7.6.

```
1.  import math
2.  import numpy as np
3.  import matplotlib.pyplot as plt
4.
5.  class projectile:
6.
7.      def __init__(self):
8.          self.g         = -9.8
9.          self.R         = 0
10.         self.H         = 0
11.         self.T         = 0
12.         self.k_list    = []
13.         self.param_str = ''
14.
15.     def get_density_fraction(self, y_in):
16.         f = math.exp ( -y_in / 7000 )
17.         return f
18.
19.     def update_acceleration(self,K, y_in, vx_in, vy_in):
20.         f = self.get_density_fraction(y_in)
21.         ax0 = -0.5*K*f*vx_in*vx_in
22.         ay0 = -0.5*K*f*vy_in*vy_in + self.g
23.         return ax0, ay0
24.
25.     def run_model(self,K, V, theta):
26.
27.         vy = [];    vx = []
28.         x  = [];    y  = [];    v  = [];   t = []
29.         yi = 0;     xi = 0;     ti = 0;
30.
31.         dt        = .1
32.         theta_rad = np.deg2rad(theta)
33.         vxi       = V*math.cos(theta_rad)
34.         vyi       = V*math.sin(theta_rad)
35.
36.         while True:
37.             ax, ay = self.update_acceleration(K,yi, vxi, vyi)
38.             dy = vyi*dt + 0.5*ay*dt*dt
39.             yf = yi + dy
40.
41.             if yf < 0 :
42.                 break
43.             else:
44.                 dx = dt*vxi + 0.5*ax*dt*dt
45.                 xf = xi + dx
46.                 x.append(xi)
47.                 y.append(yi)
48.
49.                 vyf = vyi + ay*dt
50.                 vxf = vxi + ax*dt
51.
52.                 vy.append(vyi)
53.                 vx.append(vxi)
54.                 v.append(math.sqrt(vxi*vxi + vyi*vyi))
55.                 t.append(ti)
56.                 tf = ti+dt
57.
58.                 vyi = vyf       # done this way to be explicit about
59.                 vxi = vxf       # initial/final settings
60.
61.                 xi  = xf;   yi  = yf;   ti  = tf
62.         self.R = int(xi)
63.         self.H = int(max(y))
64.         self.T = int(ti)
65.         return x, y, t, v
66.
67.
68.     def plot_xy(self, K, V, theta):
69.         p.x, p.y, p.t, p.v = p.run_model(K, V, theta)
70.         title_str="Projectile Height vs Distance " + self.param_str
71.
72.         plt.title(title_str)
73.         plt.plot(p.x,p.y)
74.
75.     def plot_vt(self, K, V, theta):
```

FIGURE 7.6    Class **projectile**.

(*Continued*)

```
76.                 p.x, p.y, p.t, p.v = p.run_model(K, V, theta)
77.                 plt.plot(p.t, p.v)
78.                 title_str="Projectile Speed vs Time " + self.param_str
79.                 plt.title(title_str)
80.                 plt.xlabel("Time (s)")
81.                 plt.ylabel("Speed (m/s)")
82.
83.         def plot_vx(self, K, V, theta):
84.                 p.x, p.y, p.t, p.v = p.run_model(K, V, theta)
85.                 plt.plot(p.x, p.v,label=str(K))
86.                 plt.legend()
87.                 title_str="Projectile Speed vs Distance "+ self.param_str
88.                 plt.title(title_str)
89.                 plt.xlabel("Distance (m)")
90.                 plt.ylabel("Speed (m/s)")
91.
92.         def run_K_list(self, V, theta, k_list):
93.                 V_str= f"{V: >4d}"
94.                 theta_str = f"{theta: >2d}"
95.                 self.param_str = '[' + V_str + 'm/s, ' + theta_str + ' deg.]'
96.                 for k in k_list:
97.                     p.x, p.y, p.t, p.v = p.run_model(k, V, theta)
98.
99.                     K_str= f"{k: >.6f}"
100.                    R_str= f"{p.R: >6d}"
101.                    H_str= f"{p.H: >5d}"
102.                    T_str= f"{p.T: >3d}"
103.
104.                    print('K=' + K_str + ', V=' + V_str + ', theta=' + theta_str +
105.                          ', R=' + R_str + ', H=' + H_str + ', T='    + T_str)
106.
107.        def plot_results(self):
108.                mydpi=120
109.                fig = plt.figure(figsize=(1800/mydpi,1800/mydpi),dpi=mydpi)
110.
111.                plt.subplot(3, 1, 1)
112.                for K in self.k_list:
113.                    p.plot_xy(K,840, 45)
114.
115.                plt.subplot(3, 1, 2)
116.                for K in self.k_list:
117.                    p.plot_vt(K,840, 45)
118.
119.                plt.subplot(3, 1, 3)
120.                for K in self.k_list:
121.                    p.plot_vx(K,840, 45)
122.
123.                plt.show()
124.                plt.savefig('./Fig 7.5.jpg', dpi=mydpi)
125.
126. if __name__ == '__main__':
127.     p = projectile()
128.
129.     p.k_list= [0, .000001, .00001, .00036, .001]
130.
131.     p.plot_results()
132.
```

FIGURE 7.6 (CONTINUED)  Class **projectile**.

Two important methods are shown, **get_density_fraction** for calculating the density fraction (**f**) at a given height (see line 15) and **update_acceleration** (line 19) for updating the x and y components of acceleration. The density fraction is based on a common exponential decay with a scale height of 7000m, so when y = 0, **f** = 1. The accelerations are based on the model described earlier where the x and y velocities are used, along with **f**, to get the current acceleration components. **K** is a parameter specified when running a model.

The **run_model** function (line 25) is used to run a model for a specified **K**, and launch configuration **V**, and **theta**. A time interval of 0.1s (line 31) was used based on testing and knowing that time in flight values between 30s and 180s were typically encountered. The function keeps looping until **y** turns negative, corresponding to the projectile returning back to ground. While not needed, variables for initial and final values in each interval were explicitly tracked for clarity and Python lists for **x, y, v,** and **t** built and returned as the function output. Range (**R**), max height (**H**), and time in flight (**T**) are tracked as class variables.

The basic idea here is that for each timestep, there are initial quantities for **x, y, $v_x$, $v_y$, $a_x$, $a_y$,** and **t**. The accelerations ($a_x$ and $a_y$) are updated based on the timestep's initial height and velocities (line 37), and these are then used to calculate the height and distance changes that occur during the interval (time slice). From these, the ending quantities (**xf, yf, vxf,** and **vyf**) are calculated, and in preparation for the next interval, the initial values are set to final values (lines 58–59). Important trajectory metrics such as **R, T,** and **H** are updated and saved.

Methods are included in the projectile class to support plotting various variable combinations, and also to run a list of models for different **K** values. Note that **run_K_list()** simply writes text to the console and returns no variables.

To support running a selection of models, to investigate the effects of using different launch and **K** parameters, models are run by iterating over a list of **K** values (**k_list**) as shown on line 129.

Finally a graphical output is generated by a routine **plot_results()** (line 107) that builds a three-panel display, where each panel is created by its own plot function.

## SUMMARY

In this chapter, we took the traditional projectile problem taught in introductory college physics courses and modified the standard equations of motion by incorporating the effects of drag. These effects were modeled by adding additional accelerations for the x and y directions, that depended on air density and on a catch-all parameter K that was tuned to match observations. Our results were surprisingly good for a system (the German 88) that includes both subsonic and supersonic flight. Our model computer code is quite general in application and could be easily applied to other projectiles – as long as the K parameter is calibrated by having the model match accepted performance metrics.

In the next chapter, we will again use a time-slicing approach, but this time we will be using dynamics to explore the problem of launching rockets into orbit, and why we need multi-stage rockets.

## CHAPTER 8

# Rocket Launches

W<small>HEN EXPLORING PARTICLE TRAJECTORIES</small> in the previous chapter, we used parameters to modify performance to match observations – quite successfully. Part of the reason we did this is we did not use dynamical equations (i.e., use equations involving forces and masses); our projectiles had no mass – it was simply an idealized particle moving under the influence of accelerations caused by drag and gravity. To model drag forces, we simply parameterized changes to the associated accelerations. With rocket launches, we are interested in the mass present, since, as it turns out, rocket mass changes substantially throughout a launch, and we encounter an interesting effect, where the fuel mass needed to launch a payload must also support the weight of the fuel itself. In some cases, simply adding more fuel does not work, since that just adds weight to the rocket, undermining the additional fuel's hoped-for benefits.

Some of the limitations encountered arise from real-world technical limitations – a given rocket engine has limited thrust – the engineers do their best to get the greatest thrust from the engine's fuel burn in a controlled stable manner, without melting the rocket. Yes, some rocket engines will be more powerful than others, but they will also be heavier, and demand more fuel, simply to lift themselves off the ground.

In this chapter, we will explore the physics of rocket launches and address questions like how high a single rocket can go and what payload it can carry. Note, when launching a payload, we care about the payload's mass and final speed. If we want a satellite to stay in an orbit of a given height, it must have a speed appropriate to that; any less means it will fall

DOI: 10.1201/9781003600046-8

lower; any greater and it will move further out. For simplicity, and without loss of generality, we will only consider circular orbits and assume the rocket is maneuvered so as to inject the payload tangentially into the orbit at the delivery speed.

## ROCKET LAUNCH DYNAMICS

When exploring rocket launches, we will assume the engine thrust is constant throughout the fuel burn phase. This does not mean the acceleration will be constant, since F = ma (force is mass times acceleration), because of fuel burn, the mass changes and decreases continuously, so the rocket is continuously accelerating. Also, the gravitational acceleration is weakening as height is gained. We will ignore drag, although it could be managed through a similar parameterization as was done for our projectile studies.

We will again use a time-slicing approach to solving this problem. At any point, the net acceleration will be calculated based on the current mass and height and the effects of the thrust and gravity; the mass will be reduced based on the flow rate and time-slice interval, and the changes in height and velocity updated for use in the subsequent time interval. To initialize the problem, we need to specify the rocket, fuel, and payload masses, the thrust, and fuel burn time.

The initialization needs the following parameters:

1. The initial MR, MF, and MP – the rocket, fuel, and payload masses

2. F, the rocket engine thrust

3. The fuel burn time Tb

4. Estimated fuel burn rate: r = MF/T

To time-slice the launch, we need to set the number of increments during the fuel burn stage, N. This means a time-slice is $\Delta t$ = Tb/N seconds long. Now we can estimate the variables at the beginning of the $i^{th}$ time-slice from those already calculated in the previous slice (i–1):

$$\text{Mass: } M(i) = M(i-1) - \Delta t * r \tag{8.1}$$

$$\text{Height gain: } \Delta h = v(i) * \Delta t \tag{8.2}$$

$$\text{Acceleration: } a(i) = M(i-1) / F + g(i) \tag{8.3}$$

$$\text{Height: } h(i) = h(i+1) + \Delta h \tag{8.4}$$

$$\text{G acceleration: } g(i) = g(h(i)) = 9.86^* \left(Re/(Re+h)\right)^2 \tag{8.5}$$

$$\text{Velocity: } v(i) = v(i-1) + a(i) * \Delta t \tag{8.6}$$

When calculating the gravitational acceleration, we use the fact it decreases as the square of the distance from the Earth's center, Re is the Earth's radius, and 9.81 is the magnitude of the gravitational acceleration at the Earth's surface.

After N steps, we reach the end of the launch, with h and v, dictating whether the payload can be satisfactorily placed into the desired orbit.

While this works in the vertical direction, in practice, space missions often happen in multiple stages where the first stage is used to enter a parking orbit during which fuel burn is used to accelerate the spacecraft, and then another stage used to climb to a higher one – perhaps for a mission to the Moon. For this reason, we will explore models with multiple rocket stages and include an option to specify a maximum height during each stage, and once that has been reached, we will assume the rocket has been rotated so it is simply accelerating at constant height.

To calculate the in-orbit acceleration, we simply calculate the acceleration ignoring g, since the rocket is moving perpendicular to the gravitational field.

## LAUNCH SIMULATIONS

We would like to explore questions like the following:

How high can a single stage rocket go?

What is the payload for a Saturn V three-stage rocket if it needs to place it in a Low Earth Orbit (LEO)?

What is the maximum payload if the Saturn V must place it in a Trans Lunar Injection orbit?

As we shall see, because there are limitations on how much thrust a rocket engine can generate, a single stage rocket cannot get much beyond LEO. For the single stage model, we will use the first-stage engine from a Saturn V and see how high and how fast it can go. The speed is a critical factor here, because near the Earth's surface, the escape velocity is about 11 km/s. This means, a payload/satellite moving at less than this speed cannot escape the Earth's gravitational field. It's not simply a problem of adding extra fuel, as this can actually make things worse as more fuel is consumed to support this additional weight!

With the LEO quest, we consider the configuration for an orbit of about 200 km high and also for a Lunar Mission, where the payload must be delivered into the proper trajectory at a speed of about 10 km/s. As noted previously for all missions there will be height and speed constraints, and the payload and fuel mass must be specified to match them. In addition, we also wish to explore scenarios where a spacecraft climbs to a temporary orbit where it accelerates and then departs to a position suitable for a mission to the Moon. So we will add a parameter (**Hmax**) to the model so if the height exceeds **Hmax**, the acceleration is done in-orbit, that is, gravitational acceleration is ignored, and no additional height gained.

Our program consists of a Python class (**rocket**) with the complete code listed under the Programming Notes, later in the chapter.

Modelling a multi-stage launch consists of getting the rocket stage parameters, and calculating the height, velocity, acceleration, and mass for each time step in each stage. Note each stage has a different gross mass, fuel capacity, and thrust. Text results are printed to the console, and a chart of the height, speed, and remaining mass is produced. Between stages, the total mass used to initialize the following stage is decremented by the burned-out stage's gross mass. Remember, the total mass for any stage burn is the sum of the remaining stage's gross masses and the cargo.

With our rocket class design, we can easily run sequences of launches, for single or multiple stages, for various payloads, with various parking orbits. Rocket stages can be modified easily by changing their thrust, engine and gross (hence fuel) masses, and other multi-stage launches added.

In Figure 8.1, we show the instruction set to use our **rocket** class to run two different model types, a single stage rocket based on the Saturn V first stage engine (**run_saturn_V_I**) to see how high it can go with a minimal payload, and then a three-stage launch (**run_saturn_V_TLI**) to see if our models show that escape velocity can be reached with a typical Lunar Landing payload.

**Hmax** is a parameter used to set the parking orbit height, that is, on reaching this height, the craft accelerates at that height. For the single-stage launch model, it only needs to be higher than fuel exhaustion for our

```
1.      r = rocket()
2.
3.      r.cargo = 140000 # LEO 170km
4.      Hmax = 5000e3
5.      r.run_saturn_V_I(Hmax)
6.      r.plot_HVM()
7.
8.      r.cargo = 50000  # TLI
9.      r.run_saturn_V_TLI()
10.     r.plot_HVM()
11.
```

FIGURE 8.1    Using our rocket class, we can run a three-stage launch for Trans Lunar Injection (cargo 50,000 kg) and also to study the performance of a single stage main engine (140,000 kg payload). Cargo is specified, and its influence on end results determined.

```
 1. Saturn V 1st Stage Performance
 2. ++++++++++++++++ STAGE 1
 3.    t       m   F(kN)  h(km)    V(m/s)      a     F/m
 4.   155  329252  34500   154.2   3474.8   78.0   104.8
 5.
 6.
 7.   Saturn V TLI: Climb to 200km, accelerate, then climb to 300km for TLI
 8. ++++++++++++++++ STAGE 1
 9.    t       m   F(kN)  h(km)    V(m/s)      a     F/m
10.   155  838252  34500    80.3   1661.1   28.6    41.2
11. ++++++++++++++++ STAGE 2
12.    t       m   F(kN)  h(km)    V(m/s)      a     F/m
13.   515  205000   4900   200.5   5493.9   23.2    23.9
14. ++++++++++++++++ STAGE 3
15.    t       m   F(kN)  h(km)    V(m/s)      a     F/m
16.  1015   60000   1033   309.8  10191.2   16.9    17.2
17.
```

FIGURE 8.2    Model results from the three-stage and single-stage models with 50,000kg and 1,000kg payloads, respectively. Column 'a' shows the net acceleration when gravity is included, while F/m shows the acceleration acting on the remaining total mass.

purposes, and setting it to 500km will exceed single-stage limits and ensure no parking orbit phase is used.

The results of the simulations are printed to the console and shown in Figure 8.2 where payloads (cargo) of 50000 kg and 1000 kg were used for the 3- and 1- stage models. Even with such a small payload, the single-stage rocket can barely make it into LEO. Earth's gravity is just too strong, and the thrust/energy provided by current engine technology is insufficient to go higher. On the other hand, the three-stage rocket model can accelerate the 50,000 kg payload up near the needed escape velocity.

The launch profiles are shown in Figures 8.4 and 8.5. In all cases, the remaining/final mass is a very small fraction of the initial mass which consisted of the fuel mass consumed and the weight of the stages that were discarded. Smoother curves would probably result if a phasing in of the next stage was begun before full exhaustion of the previous.

Even though our models are pretty simple, they are successful in revealing the underlying physics of rocket launches. With current technology

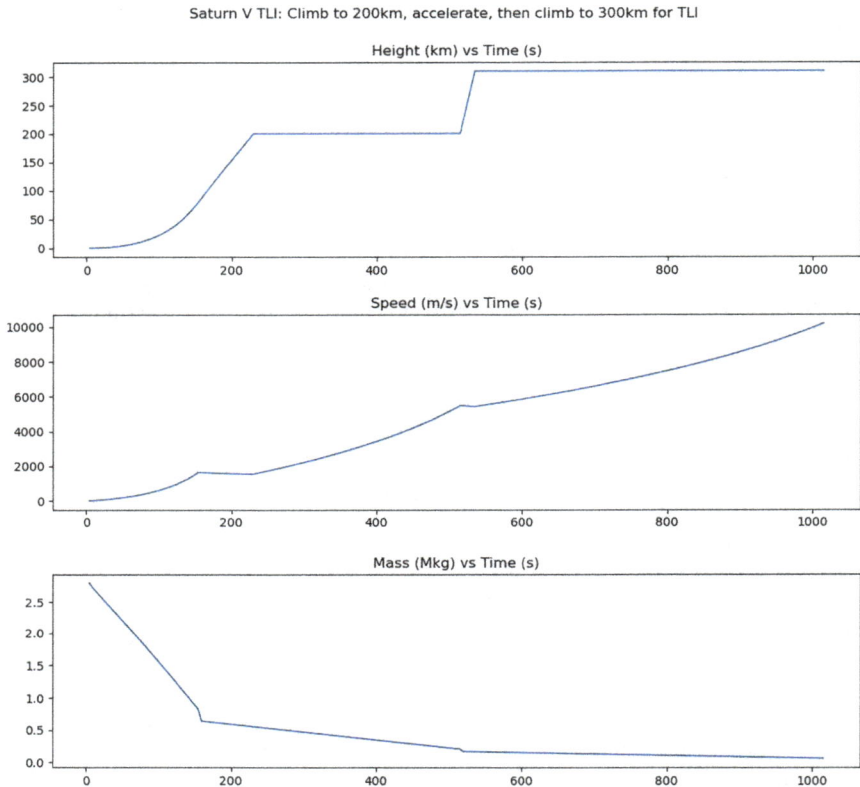

FIGURE 8.3   For this three-stage model with a 50000 kg payload, the rocket climbs to 200 km, accelerates in orbit, and then climbs to 300 km to accelerate up the escape velocity. The mass at a given time is the total number of remaining stage engine masses, payload, and remaining fuel.

placing limits on engine thrusts, they demonstrate there is no way a single stage engine can attain escape velocity or go beyond LEO. The solution is seen to be to use multi-stage rockets where rocket mass can be reduced by jettisoning engines as their fuel is used up. In many ways, this is a remarkable result in the sense that since fuel mass is the bulk of the overall mass, that it should make such a difference to be able to jettison the engine masses (Figure 8.3).

There are lots of possible experiments students could undertake with models like these, such as future engine thrust improvements, and perhaps investigating what g would make escaping LEO impossible. In addition, it would also be possible to explore drag effects in a manner similar

Saturn V 1st Stage Performance

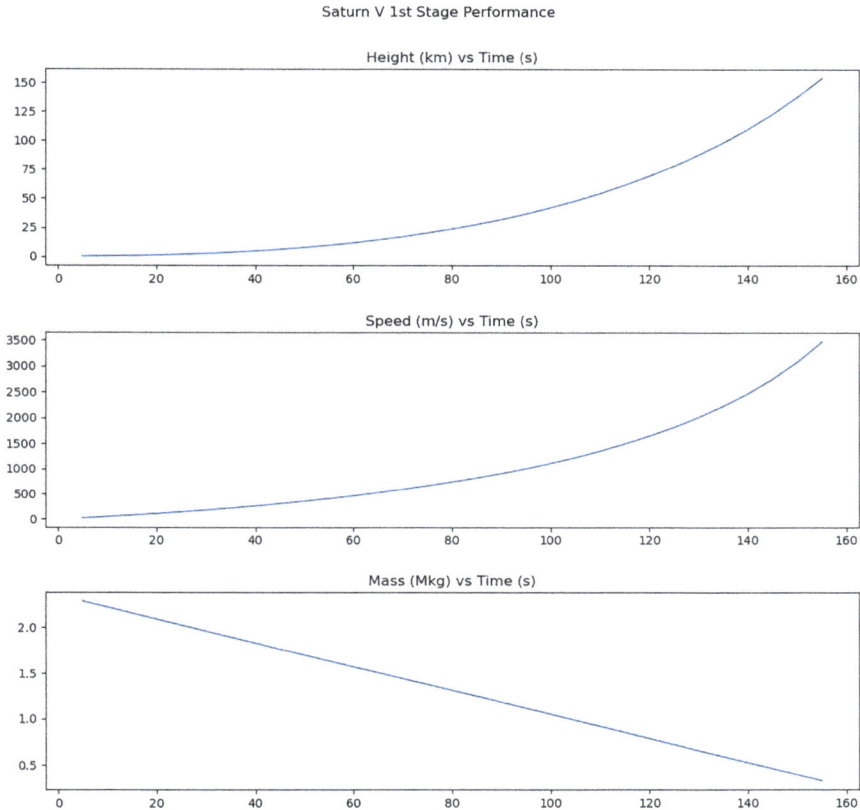

FIGURE 8.4 Single stage model. Note the maximum height is very low and the final velocity is well below that needed to escape the Earth's gravity. Mass is the engine mass, fuel remaining, and payload.

to that used for our projectile models. And it would also be very easy to add additional rocket designs/specifications for class projects.

A particularly tricky class of scenarios involves solving a launch design problem such as finding what fuel mass is required to inject a payload into an orbit at a particular height and velocity. The challenge here is that the velocity must match that for the orbit. Students might find they can only get to a lower orbit at the desired speed, but that by adding additional fuel, they reach the desired orbit but at the wrong speed – fundamentally because to gain the additional height, they had to add additional fuel to carry the additional fuel! In cases like this, there might not be any solution other than to add additional payload, that is, ballast, and having to iterate between ballast and fuel load is a fun challenge.

## CLASS ROCKET PROGRAMMING NOTES

In this section, we include the full Python code used for our rocket launch models, shown in Figure 8.5. The structure is straightforward. There are initialization functions for both the class itself (**__init__**) and for the

```python
1.  import matplotlib.pyplot as plt
2.
3.  class rocket:
4.      def __init__(self):
5.          self.dt  = 5
6.          self.g0  = -9.81
7.          self.Re  = 6357e3
8.
9.
10.         self.model_descr = ""
11.
12.     def initialize_model(self):            ########## Initialize model run
13.         self.h = 0
14.         self.v = 0
15.         self.t = 0
16.         self.H  = []
17.         self.V  = []
18.         self.T  = []
19.         self.M  = []
20.
21.     def plot_HVM(self,img_fname):               ########## PLOT RESULTS
22.         mydpi=100
23.         fig = plt.figure(figsize=(1200/mydpi,1200/mydpi),dpi=mydpi)
24.         fig.subplots_adjust(wspace=.1, hspace=.5)
25.         fig.suptitle(self.model_descr)
26.         plt.subplot(3,1,1)
27.         plt.title('Height (km) vs Time (s)')
28.         plt.plot(self.T, self.H)
29.         plt.subplot(3,1,2)
30.         plt.title('Speed (m/s) vs Time (s)')
31.         plt.plot(self.T, self.V)
32.         plt.subplot(3,1,3)
33.         plt.title('Mass (Mkg) vs Time (s)')
34.         plt.plot(self.T, self.M)
35.         plt.show()
36.         plt.savefig(img_fname, dpi=mydpi)
37.
38.
39.     def get_stage_pars(self,stage):         ########## GET ROCKET ENGINE PARAMETERS
40.         rdict = {}
41.         if stage == 'saturn5_I':
42.             Mr  = 137000
43.             Mgr = 2214000
44.             Mf  = Mgr - Mr
45.             tB  = 168 -9
46.             F   = 34500000
47.             rB  = -Mf/tB
48.         elif stage == 'saturn5_II':
49.             Mr  = 36000
50.             Mgr = 480000
51.             Mf  = Mgr - Mr
52.             tB  = 360
53.             F   = 4900000
54.             rB  = -Mf/tB
55.         elif stage == 'saturn5_III':
56.             Mr  = 10000
57.             Mgr = 119000
58.             Mf  = Mgr - Mr
59.             tB  = 165+335
60.             F   = 1033000
61.             rB  = -Mf/tB
62.
63.         rdict['Mengine'] = Mr
64.         rdict['Mgross']  = Mgr
```

FIGURE 8.5   Class rocket.                                                  (*Continued*)

```
65.            rdict['Mfuel']   = Mf
66.            rdict['tBurn']   = tB
67.            rdict['Thrust']  = F
68.            rdict['burnRate']= rB
69.
70.            return rdict
71.
72.    def print_status(self,t,m,F,h,v,a,a0):  ########## PRINT RESULTS TO CONSOLE
73.
74.            print('{0: 5d} {1: 7.0f} {2: 4.0f} {3: 7.1f} {4: 8.1f} {5: 6.1f} {6: 6.1f}'\
75.                  .format(t,m,F/1000,h/1000,v,a,a0))
76.
77.    def run_stage(self, m,Hmax,pars):        ########## RUN ROCKET ENGINE STAGE
78.            dt = self.dt                     # Hmax is parking orbit height
79.            g0 = self.g0                     # On reaching Hmax, the craft accelerates
80.            Re = self.Re                     # in-orbit, at constant height
81.            hi = self.h
82.            vi = self.v
83.            t = self.t
84.
85.            mi = m
86.            [Mr, Mgr, Mf, tB, F, rB] = list(pars.values())
87.            N = int(tB/dt)
88.            print('     t       m    F(kN)  h(km)    V(m/s)       a    F/m')
89.            for i in range(N):
90.                Ri = self.Re + hi
91.                g = g0*(Re/Ri)**2
92.
93.                mf = mi + rB*dt
94.                if hi < Hmax:
95.                    g = g0*(Re/Ri)**2
96.                    a = g + F/mi
97.                    vf = vi + a*dt
98.                    dh = 0.5*(vf+vi)*dt
99.                    hf = hi + dh
100.                   hi = hf
101.               else: # accelerate at constant h (ignoring any dh from mvr cons.)
102.                   a = F/mi
103.                   vf = vi + a*dt
104.
105.               t += dt
106.               vi = vf
107.               mi = mf
108.
109.               self.T.append(t)
110.               self.H.append(hf/1000)
111.               self.V.append(vf)
112.               self.M.append(mf/1000000.)
113.
114.           self.print_status(t,mf,F,hf,vf,a,F/mf)
115.
116.           self.h = hf
117.           self.v = vf
118.           self.t = t
119.
120.
121.    def run_saturn_V_I(self,Hmax):          ########## RUN Saturn V FIRST STAGE MODEL
122.           self.model_descr="Saturn V 1st Stage Performance"
123.           self.initialize_model()
124.           s1 = r.get_stage_pars('saturn5_I')
125.           mTot = self.cargo + s1['Mgross']
126.           print('\n\n',self.model_descr)
127.           print('++++++++++++++ STAGE 1')
128.           self.run_stage(mTot,Hmax,s1)
129.
130.
131.    def run_saturn_V_TLI(self):             ########## RUN Saturn V TLI MODEL
132.           self.model_descr=\
133.               "Saturn V TLI: Climb to 200km, accelerate, then climb to 300km for TLI"
134.           s1 = r.get_stage_pars('saturn5_I')
135.           s2 = r.get_stage_pars('saturn5_II')
136.           s3 = r.get_stage_pars('saturn5_III')
137.
138.           mTot = self.cargo + s1['Mgross'] + s2['Mgross'] + s3['Mgross']
139.           self.initialize_model()
140.           Hmax = 200e3
```

FIGURE 8.5 (CONTINUED)  Class rocket.                    (*Continued*)

```
141.                print('\n\n',self.model_descr)
142.                print('++++++++++++++ STAGE 1')
143.                self.run_stage(mTot,Hmax,s1)
144.
145.                # stage 1 separation
146.                mTot -= s1['Mgross']
147.
148.                print('++++++++++++++ STAGE 2')
149.                self.run_stage(mTot,Hmax,s2)
150.
151.                # stage 2 separation
152.                mTot -= s2['Mgross']
153.
154.                Hmax = 300e3
155.                print('++++++++++++++ STAGE 3')
156.                self.run_stage(mTot,Hmax,s3)
157.
158.
159. ################################## END OF CLASS DEFINITIONS ##############
160.
161. if __name__ == "__main__":
162.
163.     r = rocket()
164.
165.     r.cargo = 140000 # LEO 170km
166.     Hmax = 500e3
167.     r.run_saturn_V_I(Hmax)
168.     r.plot_HVM('SingleStage.jpg')
169.
170.     r.cargo = 50000  # TLI
171.     r.run_saturn_V_TLI()
172.     r.plot_HVM('TLI.jpg')
173.
```

FIGURE 8.5 (CONTINUED)   Class rocket.

start of each model run (**initialize_model**). For output, there is a method to plot height, velocity, and mass (**plot_HVM**), and one for text values (**print_status**). Two models are included to explore either a single stage launch (**run_saturn_V_I**) or a three-stage launch (**run_saturn_V_TLI**), with a function to set the parameters for the rocket stages (**get_stage_pars**) and implementing a given stage's burn (**run_stage**).

The **get_stage_pars** method returns a dictionary of properties for the requested stage. Our models are based on posted values for the Saturn V rocket. Note, other designs can be easily added and existing ones modified. To change the mass of fuel, simply change the gross mass for a stage.

Our single-stage model simply uses the parameters for stage I of a three-stage Saturn V system, and while this might not be used in practice, they are physically reasonable quantities and convenient to use.

For a specified engine/stage, the simulation is time-sliced in intervals of **dt** with the acceleration calculated in each time-slice based on the current total mass (line 96). The mass is reduced in each time slice; therefore, since we assume the thrust is fixed for a given engine, the acceleration from thrust is **F/m** and the net acceleration is **F/m + g**, where **g** is calculated for the height at line 95 (really the distance from the Earth's center); for in-orbit acceleration where the motion is perpendicular to **g**, **g** is ignored and we also assume the height is not changing from conservation of angular momentum.

During each time-slice, properties like acceleration, height, mass, and speed are updated, based on initial values, and the resulting ones are set as the initial values for the following time-slice (lines 94–115).

## SUMMARY

In this chapter, we used time-slicing to analyze the rocket launch model in which the changing mass caused by fuel burn could be taken into account. While there are analytic solutions, it becomes significantly more intractable if changing gravitational effects with height are also considered. Because our models are based on very basic physics, they are appropriate for young students, and their focus can be directed toward investigating various scenarios with different height and payload constraints. Our models demonstrated why single stage rockets cannot be used to escape Earth's gravitational field, and how this could be done using multi-stage ones. The models could be easily extended and customized to match the many possible designs.

## CHAPTER 9

# Building a Star Catalog from an Image

STRONOMERS TEND TO TAKE lots of images because images allow information about an extended object, or of multiple objects to be collected all at once. It wasn't always that way. In the past, photometers or spectroscopes would be used on individual targets, but newer technology supports more efficient wider field views.

What does one do with an image? We might be interested, for example, in the brightness of individual targets (stars/asteroids/galaxies) or in their positions. So how do we get from a digital image to an inventory of image targets? What would such an inventory look like? At a minimum, we'd like to know where each object is located, how bright it is, and its size. For a really detailed analysis, we might care about its shape.

With our eyes, we can easily pick out the stars, but we need a mechanism where we can have the computer do this for us. In this chapter, we will describe a basic solution that will detect stars based on a threshold intensity, which will create a list of regions containing pixels with intensities greater than the threshold. These detected regions will generally be our stars or targets of interest, and by counting the number of pixels in a region, we can estimate its size, and by adding the pixels, the region's flux. Also, we can find the center of mass of each region which will correspond to the region's center – a better estimate of the star's position than that of the brightest pixel.

DOI: 10.1201/9781003600046-9

For brighter objects, they will be well above the threshold level and their properties reasonably detected; very faint objects however might be missed or have their sizes and fluxes underestimated. This is just the nature of things. One can try reducing the threshold, but that might introduce false detections into our catalog from artifacts such as scattered light or gradients across the image; it's a judgment call where to set the threshold. For our purposes, we will use a threshold that picks out the brightest stars in our test image and the reader can adjust the strategy to suit their needs.

The task we are facing consists of the following elements:

- Read in an image

- Identify targets (stars/asteroids)

- Estimate the position and fluxes for each target

- Create a catalog of the targets found

With catalogs like these, we could compare targets for catalogs created at different times to see if any targets moved or changed brightness, in support of asteroid or supernova or variable star studies. In addition, we might compare them to highly accurate star catalogs in order to assign astronomical coordinates such as RA and DEC to our targets. Or we could use target position as input to more detailed photometry functions (beyond simple flux counts of pixels exceeding a particular threshold) such as single- and multi-iris photometry. In our case, we developed a class called 'imcat' that can analyze an image, generate a catalog, and print out a chart of the detected objects which can be compared with the input image. Script test_imcat.py, shown in Figure 9.1 demonstrates how it was applied to an image of the T CrB region. (T CrB is a recurring nova that was predicted to go nova around the time of writing, and we imaged it using a smart scope.)

The rendered star sizes in the plotted star chart uses a scaling factor of 4000 (see Figure 9.1 line 14) determined through trial and error, and based on star fluxes. Smaller stars are achieved using larger scale factors.

The image processed in this example is shown in Figure 9.2 and the resulting catalog's chart is shown in Figure 9.3.

This code assumes there is a flat field image called flat.jpg in the same directory. The target image was taken using a Seestar 50 on May 11, 2024.

```
 1. imc              = imcat.imcat()
 2. imc.proj_dir     = './'
 3. imc.fname        = 'T_CrB_20240511'
 4. imc.im_source    = 'Seestar50'
 5. imc.get_flattened_img()
 6.
 7. # set detection threshold based on image mean value
 8. imc.T = imc.img.mean()*2.
 9.
10. imc.build_img_catalog()
11. imc.write_star_cat
12.
13. # plot 300 brightest catalog entries with scale factor 4000
14. imc.plot_catalog_xy_coordinates(300,4000)
15.
```

FIGURE 9.1   A simple code, **test_imcat.py** to test the **imcat** class on a jpg image. The code generates a star catalog from the image and produces a chart of the 300 brightest stars.

FIGURE 9.2   The jpg image of T CrB (center) used to build the catalog. The field is about 1.2 x .75 degrees. Note the image label, which is flagged to be trimmed using the my_ source='Seestar50' entry at line 15 of Figure 9.1.

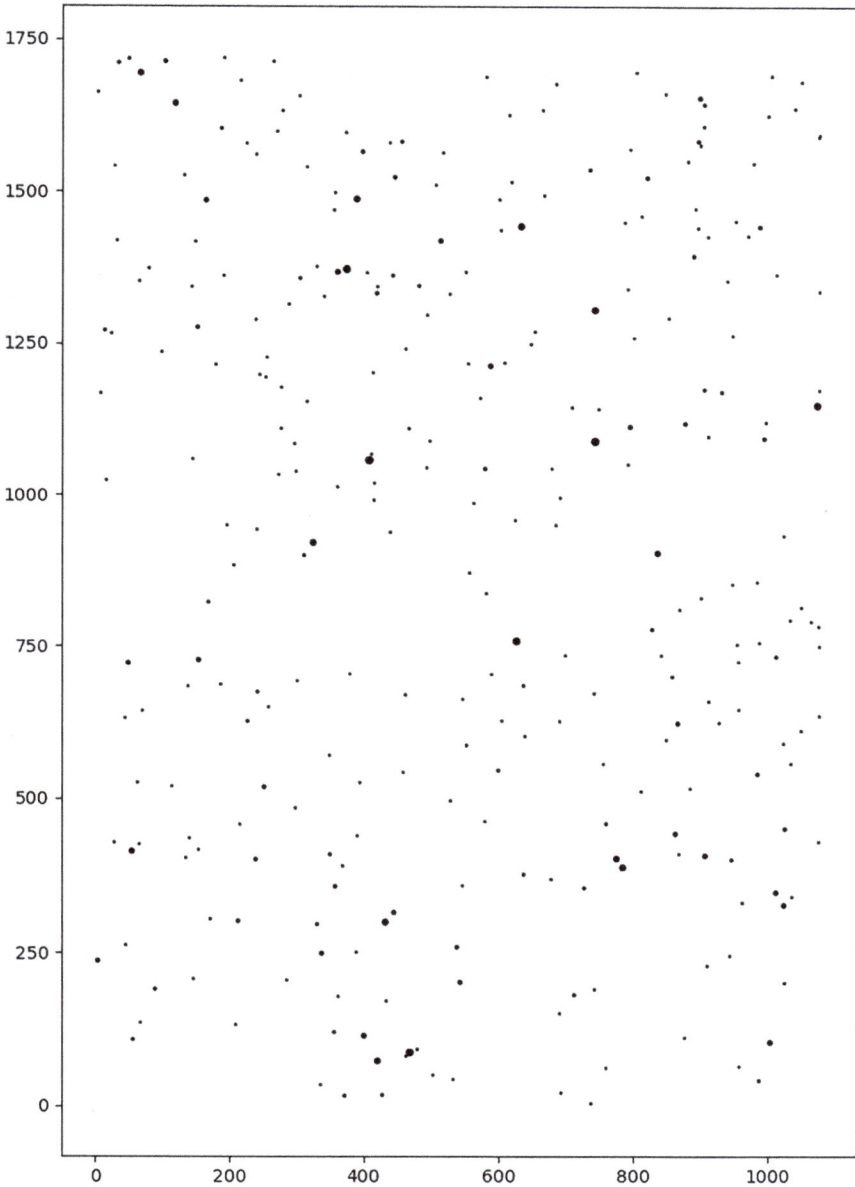

FIGURE 9.3 A chart produced from the image catalog suggests that the catalog building was quite successful. Here the stars are circles, whose sizes are scaled to catalog-integrated fluxes.

The Seestar 50 image is used here because smart scopes like it are becoming very popular. However, their jpg images can have footer text, and by setting my **im_source** parameter to Seestar50, we can trim out the image footer with the text. If not using the label feature with Seestar 50 or an image without a label just leave **im_source** empty.

The code in Figure 9.1 is very simple – all the difficult stuff is hidden away in the **imcat** class. The detection threshold is set as a multiple of the average intensity. In this instance we used 2.0 through trial and error (see Figure 9.1 line 8), and we found that increasing this much more would result in unacceptably fewer detections, while reducing it closer to 1.0 would produce an overwhelming and unmanageable/unwanted number. When producing the chart, the function was passed a value to set the number of stars to use (line 14). Since the catalogs are sorted with the brightest targets first, in the example only the brightest 300 targets are drawn.

We will now describe the **imcat** class in detail.

## CLASS IMCAT PROGRAMMING NOTES

The **imcat** class uses functions that handle initialization, image I/O, star detection, and catalog construction and output, and many of these are organized into other classes: **imcat_io**, **imcat_pixels**, **imcat_cat**, used by **imcat**.

### Class imcat_io

Image I/O is supported by functions in class **imcat_io** shown in Figure 9.4.

If an image flat is present, it can be used to flatten the main image being cataloged. The jpg flat file is read in by the **read_flat** function (line 8), converted into a grayscale (line 11), and normalized so the maximum pixel intensity is unity (lines 13–14).

The image to be cataloged is read in as a jpg file and converted to grayscale and saved to **self.img** (lines 16–21).

Image flattening is done by function **apply_flat**, by dividing each pixel in **self.img** by the same pixel in **self.flat** which produces a new flattened image **self.imgf**.

Because it is useful to work with a subset of an image, function **set_subim** can select a sub-image of **self.img** (line 34) and set appropriately sized utility images/arrays (**self.flat**, **self.ims**, **self.imgf**). Each pixel in the **self.ims** array holds the star number associated with the pixel; this allows us to identify neighboring stars when looking at a given pixel.

```
 1. from PIL import Image, ImageOps
 2. import numpy as np
 3.
 4. class imcat_io:
 5.     def __init__(self):
 6.         pass
 7.
 8.     def read_flat(self):
 9.         file        = self.proj_dir+'flat.jpg'
10.         im_in       = Image.open(file)
11.         im_in_gray  = ImageOps.grayscale(im_in)
12.         flat0       = np.asarray(im_in_gray)
13.         fmax        = flat0.max()
14.         self.flat   = flat0/fmax
15.
16.     def read_img(self):
17.         self.img_file   = self.fname + '.jpg'
18.         file            = self.proj_dir+self.img_file
19.         im_in           = Image.open(file)
20.         im_in_gray      = ImageOps.grayscale(im_in)
21.         self.img        = np.asarray(im_in_gray)
22.
23.     def apply_flat(self):
24.         [i1,j1] = self.subim_dims
25.         for i in range(i1):
26.             for j in range(j1):
27.                 self.imgf[i,j] = self.img[i,j]/self.flat[i,j]
28.
29.     def set_subim(self):
30.         (i1,j1)     = self.img.shape
31.         if self.im_source == 'Seestar50':  # ignore labels
32.             i1 -= 200
33.         self.subim_dims = [i1,j1]
34.         self.subim  = self.img[0:i1,0:j1] #just accept full image
35.         self.ilen   = i1
36.         self.jlen   = j1
37.         self.Imin   = self.subim.min()
38.         self.Imax   = self.subim.max()
39.         self.ims    = np.zeros(shape=(self.ilen, self.jlen), dtype=int)
40.         self.flat   = np.zeros(shape=(self.ilen, self.jlen), dtype=int)
41.         self.imgf   = np.zeros(shape=(self.ilen, self.jlen), dtype=float)
42.
43.     def get_flattened_img(self):
44.         self.read_img()
45.         self.set_subim()
46.         self.read_flat()
47.         self.apply_flat()
48.         self.img = self.imgf
49.
```

FIGURE 9.4   Class **imcat_io** has functions to read in and manipulate images used by **imcat**.

Function **get_flattened_image** combines the other utility functions into a single utility to read in, resize, and flatten the target image.

## Class imcat_pixels

This class holds the four functions used to examine each pixel and determine which star it is a member of and is shown in Figure 9.5.

Because only pixels with intensities greater than the threshold are of interest, a list (**self.pixlist**) of those pixels' coordinates is constructed using function **build_pixlist()**, and this list is normally considerably smaller than the number of pixels in the image. Working with **self.pixlist** is more efficient than with **self.img**.

```
 1. class imcat_pixels:
 2.     def __init__(self):
 3.         pass
 4.
 5.     def build_pixlist(self):
 6.         im0 = self.img
 7.         T = self.T
 8.         for i in range(1,self.ilen-1):
 9.             for j in range(1,self.jlen-1):
10.                 if im0[i,j] > T:
11.                     self.pixlist.append([i,j])
12.
13.     def find_stars(self):
14.         for pix in self.pixlist:
15.             [i,j] = pix
16.             adj_stars = self.find_nearby_stars(i,j)
17.
18.             if len(adj_stars) == 0:
19.                 self.num_stars += 1
20.                 print("New star: ", self.num_stars, " at :",i,j)
21.                 self.star_pix_dict.update({self.num_stars:[[i,j]]})
22.                 self.ims[i,j] = self.num_stars
23.             elif len(adj_stars) == 1:
24.                 parent = adj_stars[0]
25.                 plist = self.star_pix_dict[parent]
26.                 plist = plist + [[i,j]]
27.                 self.star_pix_dict[parent] = plist
28.                 self.ims[i,j] = parent
29.             else:
30.                 #print('*** ADJACENT STARS:', adj_stars)
31.                 parent = self.merge_adjacent_stars(adj_stars)
32.                 plist = self.star_pix_dict[parent]
33.                 plist = plist + [[i,j]]
34.                 self.star_pix_dict[parent] = plist
35.                 self.ims[i,j] = parent
36.
37.     def find_nearby_stars(self, i,j):
38.         ims0 = self.ims
39.         s1 = ims0[i-1,j+1]
40.         s2 = ims0[i-1,j]
41.         s3 = ims0[i-1,j-1]
42.         s4 = ims0[i,j-1]
43.         s_list = [s1,s2,s3,s4]
44.         s_set = set(s_list)          # make unique
45.         s_list = list(s_set)
46.         if 0 in s_list:
47.             s_list.remove(0)
48.         s_list.sort()
49.         return s_list
50.
51.     def merge_adjacent_stars(self,adjacent_star_list):
52.         p = adjacent_star_list[0]       # parent is first in list
53.         if len(adjacent_star_list) > 1:
54.             print('adjacent stars: ', adjacent_star_list)
55.
56.             for s in adjacent_star_list[1:]:
57.                 plist = self.star_pix_dict[p]
58.                 slist = self.star_pix_dict[s]
59.                 plist = plist + slist
60.                 self.star_pix_dict[p] = plist
61.                 print("DELETING STAR: ", s)
62.                 for ij in self.star_pix_dict[s]:
63.                     i,j = ij
64.                     self.ims[i,j] = p
65.                 del self.star_pix_dict[s]
66.         return p
67.
```

FIGURE 9.5   Class **imcat_pixels**.

When finding stars, pixels in **self.pixlist** are studied to see if they are isolated from others or are touching one or more stars. To keep track of what star a pixel is in, **self.ims** is an array where instead of intensities, at each [i,j] the assigned star_number is stored.

A star is a dictionary entry (**self.star_pix_dict**) where the key is the star_number and the value is a list of pixel coordinates for each pixel in the star. For example **self.star_pix_dict[10]** might equal [[1,2],[23,21], [45,78]] To add a pixel at [46,78] to the star, its coordinates (in brackets) are appended to the list. The length of the list tells us the star area.

If a pixel is isolated, it is the first pixel of a new star, and a new star is created in **star_pix_dict**. If a pixel is touching a star, it is added to the star. If the pixel touches more than one star, the pixel forms a connecting bridge between them so those stars are merged, and the pixel added to the resulting star.

Merging two stars, A and B, simply means merging the coordinate list from B into that of A – where B has a higher star_number than A, changing the **self.ims** pixel values for B over to A, and deleting **self.star_pix_dict**[B].

As new stars are detected, new key: Value entries are added to **star_pix_dict** and as they are merged, unwanted entries are removed.

Finding stars therefore consists of the following steps:

1. For every pixel in **self.pixlist** (line 14).

2. Make a list of the nearby stars using self.ims (line 16) and turn the list into a set so only unique stars are kept, and then revert back to a list (lines 37–49).

3. Delete star 0 which is the default background star_number value (line 46–47).

4. If there are no adjacent stars, the pixel is a new star and a new star is created (lines 19–21), and **self.ims**[i,j] set to the star number (line 22).

5. If there is only one nearby star (the parent), the pixel is merged with it and **set.ims**[i,j] uses the parent star number (lines 24–28).

6. If there is more than one adjacent star, the pixel is linking them so they are part of the same star, and they need to be merged before the pixel is added to the result (lines 29–35).

These steps implemented using functions **build_pix_list(), find_stars(), find_nearby_stars()**, and **merge_adjacent_stars()**.

## Class imcat_cat

Once all the stars have been identified and saved in **star_pix_dict**, functions in class **imcat_cat** (shown in Figure 9.6) are used to build the final

```
 1. from operator import itemgetter
 2. import pandas as pd
 3.
 4. class imcat_cat:
 5.     def __init__(self):
 6.         pass
 7.
 8.     def build_catalog(self):
 9.         T = round(self.T,1)
10.         for k in self.star_pix_dict:
11.             ij_list = self.star_pix_dict[k]
12.             npix = len(ij_list)
13.             Itot = 0
14.             icg = 0
15.             jcg = 0
16.             for l in ij_list:
17.                 [i,j] = l
18.                 I = self.img[i,j]
19.                 Itot += I
20.                 icg += I*i
21.                 jcg += I*j
22.
23.             jcg  = round(jcg / Itot,2)
24.             icg  = round(icg / Itot,2)
25.             Itot = round(Itot,2)
26.
27.             star_data=[k,T, icg, jcg, Itot, npix]
28.             self.star_cat.append(star_data)
29.
30.         # largest Itot (column 4) first
31.         self.star_cat=sorted(self.star_cat,key = itemgetter(4), reverse=True)
32.         self.write_star_cat()
33.
34.     def write_star_cat(self):
35.         fname_cat = self.proj_dir + self.fname + '_cat.csv'
36.         f = open(fname_cat, 'w')
37.         for i in range(len(self.star_cat)):
38.             [k,T, icg, jcg, Itot, npix]=self.star_cat[i]
39.             kstr  = f"{k:5d}"
40.             Tstr  = f"{self.T: 3.0f}"
41.             istr  = f"{icg: >8.2f}"
42.             jstr  = f"{jcg: >8.2f}"
43.             Istr  = f"{Itot:>10.0f}"
44.             nstr  = f"{npix:>6d}"
45.             dstr = kstr+','+Tstr+','+istr+','+jstr+','+Istr+','+nstr+'\n'
46.             print(kstr,Tstr,istr,jstr,Istr,nstr)
47.             f.write(dstr)
48.         f.close()
49.
50.     def read_csv_into_df(self,N,fname):
51.         fpath = self.proj_dir  + fname +'_cat.csv'
52.         df    = pd.read_csv(fpath, nrows=N)
53.         df.columns = ["sid","T","i","j","Flux","Area"]
54.         return df
55.
```

FIGURE 9.6    Class **imcat_cat** holds functions needed for writing out and reading in star catalogs.

star catalog for the image. Output entries for each star consist of its star_ number (k) and the threshold used (T), its position based on its center of gravity (**icg, jcg**), the integrated flux (**Itot**), and area (**npix**).

The **build_catalog()** function loops through each star in **star_pix_list** and computes the star's properties (lines 11–28) which are formatted and written out to a csv file by the **write_star_cat** function (lines 34–48).

Even though the threshold is the same for all stars and is redundant to be stored as a column, we decided to leave it that way in case there was a future need to mix catalog information. Also, in the CSV catalogs, the star id (**star_number**) is only useful for debugging and is generally unused and harmless and could be omitted from them.

## Class imcat

All the functionality built in to the three supporting classes (**imcat_***) are used by class **imcat** (see Figure 9.7) to process an image, produce a star catalog, and generate a chart such as that shown in Figure 9.3, from the catalog.

Running the **imcat** class consists of getting a flattened image (line 59), setting a threshold (line 62), building and writing out the star catalog (lines 64–65), and using the catalog to produce a chart (line 68).

The star chart is produced by **plot_catalog_xy_coordinates** (line 30), which reads in the catalog csv file (line 31) – this way, a modified code could simply work with pre-existing catalogs without having to process an image.

The **plot_xy_catalog_coordinates()** function takes a specified number (N) of rows to control clutter, and the **scale_factor** used to control star sizes. It extracts four data columns (line 36) ([i,j] position, star id, and flux) and uses the [i, j] positions to draw the stars, and the star fluxes/intensities/brightnesses to set the charted star size using use the provided scaling parameter (shown as 4000). The scale factor was chosen through trial and error and a minimum star size is set to 1.

The functions defined in **imcat.py** are flexible in that they can be easily modified to suit your purpose, and it would be easy to wrap them in a loop to iterate over a directory of images. Very limited hardening was done for the sake of clarity, such as testing for empty return values. For limited scope projects, you can get away with this, for a while, but any code used for other than demonstrations should be hardened.

**get_flattened_img** will flatten **self.img** and set it to the result. This way, **self.img** can either be used in the original form or flattened. The **build_img_catalog** function is simply a one-step solution for cataloging an image, suitable for iterating over a directory of images.

The class also includes a section after the 'if __name__ ==' conditional to demonstrate how to use the class and produce a catalog and its chart shown in Figure 9.3.

```
1.  import matplotlib.pyplot as plt
2.  from imcat_io import imcat_io
3.  from imcat_pixels import imcat_pixels
4.  from imcat_cat import imcat_cat
5.
6.
7.  class imcat(imcat_io, imcat_pixels,imcat_cat):
8.      def __init__(self):
9.          self.im_source     = ''            # image type for trimming
10.         self.img           = []            # working image
11.         self.ims           = []            # image to track star membership
12.         self.imgf          = []            # image after flattening
13.         self.ilen          = 0             # image dimensions
14.         self.jlen          = 0
15.         self.Imax          = 0             # Intensity max
16.         self.Imin          = 0
17.         self.T             = 0             # thresholding level
18.         self.pixlist       = []            # list of pixels > T
19.         self.num_stars     = 0
20.         self.star_pix_dict = {}            # dictionary of star pixels
21.         self.star_cat      = []            # list of star properties
22.
23.     def get_ijsF_df_cols(self,df):
24.         v1      = df['i'].tolist()
25.         v2      = df['j'].tolist()
26.         v3      = df['sid'].tolist()
27.         v4      = df['Flux'].tolist()
28.         return v1,v2,v3,v4
29.
30.     def plot_catalog_xy_coordinates(self,N, scale_factor):
31.         df = self.read_csv_into_df(N,self.fname)
32.
33.         xin       = 4                      # set chart dimensions
34.         yin       = xin*self.ilen/self.jlen
35.
36.         i1,j1,s,f1 = self.get_ijsF_df_cols(df)
37.         s1 = [int(f/scale_factor)+1 for f in f1]
38.         y1 = []
39.         for i in i1:
40.             y1.append(self.ilen - i)
41.         x1          = j1
42.         fig, ax     = plt.subplots()
43.         fig.set_size_inches(xin,yin)
44.         ax.scatter(x1, y1, color = 'black', s=s1)
45.         plt.show()
46.
47.     def build_img_catalog(self):
48.         self.build_pixlist()
49.         self.find_stars()
50.         self.build_catalog()
51.
52.  if __name__ == '__main__':
53.      imc               = imcat()
54.
55.      imc.proj_dir      = './'
56.      imc.fname         = 'T_CrB_20240511'
57.      imc.im_source     = 'Seestar50'
58.
59.      imc.get_flattened_img()
60.
61.      # set detection threshold based on image mean value
62.      imc.T = imc.img.mean()*2.
63.
64.      imc.build_img_catalog()
65.      imc.write_star_cat
66.
67.      # plot 300 brightest catalog entries with size scale factor 4000
68.      imc.plot_catalog_xy_coordinates(300,4000)
69.
```

FIGURE 9.7    Class **imcat**.

## SUMMARY

In this chapter, a Python class was explored to generate a catalog of stars from an image. A catalog of detected objects can be a very powerful resource, and the coordinates of entries can be used to track changes/movement, be cross identified with regular star catalogs, and also serve as input to photometry tasks, where the simple flux counts are not good enough. The functions provided can be modified easily for other applications.

In the next chapter, we will explore how to make photometric measurements on an image which will require the user to select targets using a cursor, and selecting backgrounds so background components can be removed. By using a list of known stars and their magnitudes, the fluxes can then be estimated.

## CHAPTER 10

# Photometry

## *Measuring Object Brightness*

A STAR IN AN IMAGE looks like a blob with tapering edges, and astronomers doing precise photometry, when measuring the brightness of a star, will try and measure the flux within a certain radius of the star's center (the 'iris' or 'aperture'). (For our purposes, we will consider flux, pixel values, and pixel intensities as being the same.) As the radius is increased, a greater fraction of the star's light will be included; however, since there is always some background light and noise, the larger the radius, the greater the false signal from the background. In general, astronomers will decide on an optimal radius to use, measure the light within that radius of the star's center (which includes both star and background light), and then measure the background and subtract it from the first measurement.

This can be done in one step, where three circles are specified with the inner circle used to measure the target, and the annulus formed by the two outer ones used to estimate the background, which can then be used to remove the target's background level. This approach is called multi-aperture photometry.

An alternate approach is to use single-aperture photometry, where one circle/aperture is defined. This is centered on the target for a measurement and then placed nearby, so a background can be estimated.

We will use the simpler single-aperture method since only one aperture size need to be used, and since we will be doing things interactively, we

DOI: 10.1201/9781003600046-10

can simply choose where to measure the background instead of having to adjust the radii to exclude nearby stars with the multi-aperture approach.

Because we obviously need to include a function to estimate flux within a certain radius centered on the cursor position, the code could be modified to for multi-aperture photometry by defining three concentric apertures with radii $r_1 < r_2 < r_3$, measuring the fluxes in each, to yield $f_1$, $f_2$, $f_3$, and then using $f_3 - f_2$ to estimate the background level (or more likely, the background level per pixel, since the annulus area might be different from the inner circle's).

## DESIGN CHOICES AND SOFTWARE INSTRUCTIONS

We created a class (**imphot**) to do simple, manual photometry, with astronomical images. Our design was intended to support basic, manual photometry, even though it could be modified for automated use. Since it's primarily intended as a demonstration and to be easy to use, we wanted the user to be able to quickly measure a star and the star background; to do this for a short list of stars with known magnitudes, a magnitude calibration could be done, and then the targets of interest could be measured.

Since the user is using a cursor, an aperture circle is drawn on the image where the mouse is clicked to select the area of interest to visually show where the flux was measured.

When measuring a star, the drawn circle is white; when measuring the background, the circle is blue.

Pressing the letters 's' and 'b' select the measurement mode.

Pressing 'n' tells the code to save/use the most recent star and background measurements, and to move on to the next star or target.

A file in the project directory called star_mags.csv contains reference star magnitudes. The user must measure these stars first, after which the magnitude calibration is done. Then, any further targets will have their estimated magnitudes shown on the console.

A polynomial/linear fit is used to match reference star magnitudes with their flux, so a minimum of two reference stars is needed. More reference stars can be included, but this requires the user to make more measurements for the calibration, and in practice, adding too many reference stars can often degrade the results.

The size of the aperture/iris is set to a default of eight pixels but can be incremented/decremented by using the +/− keys. To change the aperture, select the star 's' mode and try different apertures to see which gives the best match and then proceed to making the reference star measurements.

(Note both an 's' and a 'b' measurement must be taken after changing the iris to ensure both the star and background areas/circles are the same, and the SNR is properly estimated.)

An estimate of the SNR (signal to noise ratio) is printed to the console when star and background measurements are available. The results and measurements are written on the console.

## Measurement Strategy

A general strategy for measuring target magnitudes requires measuring the reference stars so the magnitude calibration can be performed and then the targets of interest. In most cases, it will be necessary to use matplotlib's display and selection tools to zoom in and out, so as to best see the object being measured. This will not affect the measurements.

1. Find the best iris size by examining the target or a reference star:

    a.  Enter 'b' and make a background measurement

    b.  Enter 's' and make multiple star measurements of the first reference star

    c.  Adjust the iris using +/– and find the best SNR; do not change the iris for the remainder of the processing.

2. For each reference star:

    a.  Type 's' and click on the star's center,

    b.  Type 'b' and click on background,

    c.  Repeat steps a or b if desired,

    d.  Save the measurement and move to the next reference star (type 'n').

3. Measure the star and background levels for the other reference stars. After this, the magnitude scale is calibrated.

4. Measure the star and background levels of the targets of interest using the same 'process as for the reference stars (step 2) using 'n' to save and move on to the next target.

5. Results appear on the console and can be listed by typing **imc. star_list**.

Note, when making measurements, it doesn't matter whether the background is measured before the star, and either can be redone; only the latest ones are kept.

## TESTING imphot

To test and demonstrate the **imphot** class, a small program (**test_imphot.py**) was created, which uses the T CrB image used by test_imcat.py of an earlier chapter. The code is shown in Figure 10.1.

It basically consists of an instantiation of the **imphot** class (line 3), initialization part (lines 5–24), a small section (lines 19–21) to override windows system command/quick-key definitions so 's' can be entered without causing windows to change, a section to control image display and dimensions (lines 26–36), and code to handle events like keyboard entries (lines 32–34).

On startup the user is presented with a window (see Figure 10.2) showing the image of interest. The window can be zoomed in and out to make

```
 1. import matplotlib.pyplot as plt
 2. from imphot import imphot
 3. imp = imphot()
 4.
 5. imp.proj_dir        = './'
 6. imp.fname           = 'T_CrB_20240511'
 7. imp.img_file        = imp.fname + '.jpg'
 8. imp.star_mags_file = 'star_mags.csv'
 9.
10. imp.im_source = 'Seestar50'
11. imp.read_img()
12. imp.set_subim()
13. imp.read_flat()
14.
15. imp.apply_flat()
16. imp.img = imp.imgf
17. imp.read_star_list()  #read file with known star mags
18.
19. klist = plt.rcParams['keymap.save']
20. if 's' in klist:
21.     plt.rcParams['keymap.save'].remove('s')
22.
23. xin = 4                        # set chart dimensions
24. yin = xin*imp.ilen/imp.jlen
25.
26. fig, ax = plt.subplots(1,1)
27. fig.set_size_inches(xin,yin)
28. imp.ax = ax
29. imp.fig = fig
30. imp.ax.imshow(imp.img,cmap='gray')
31.
32. plt.connect('button_press_event', imp.on_click)
33. plt.connect('key_press_event'  , imp.on_press)
34. plt.connect('key_release_event' , imp.on_key_release)
35.
36. plt.show()
37.
```

FIGURE 10.1   A short program (test_imphot.py) that uses the **imphot** class for single-aperture photometry.

FIGURE 10.2    The image used for testing the code. The reference stars are marked, and T CrB is near the center.

it easier to select a star for analysis. When a star is measured, the user can zoom out and select a different one to zoom in on.

A zoomed-in view of the first reference star is shown in Figure 10.3, with the star iris (white) and a background iris (blue) while changing the star's iris to improve the SNR. Figure 10.4 shows the console output during the process.

Initially a background measurement was made using the default R = 8 pixels, and then a star measurement yielded an SNR of 292. Because the star iris appeared too large, R values of 7 and 6 were tried, with the SNR improving to 320 and 344. Repeating the background measurement using the R = iris brought the SNR up to 356. During these steps, the flux in the star iris went from 42041 down to 26338 (about a 40% drop), but the background flux went from 11150 to 3947 (about 60% drop), hence the SNR improvement.

A complete analysis using two reference stars and the target (measurements 0, 1, and 2) is shown in Figure 10.5. Reference stars 'm' and 'K' have

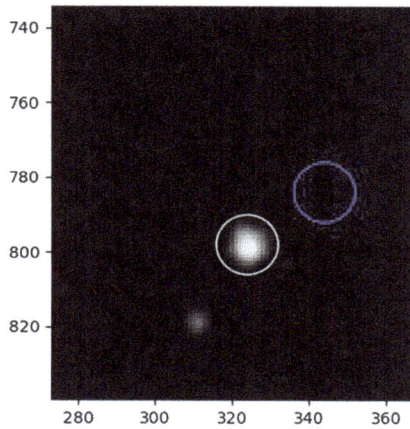

FIGURE 10.3    The view after a background was chosen. The background iris (blue) is the default 8 pixel radius and is greater than the star's iris (white) because the star's iris was decreased to improve the SNR. Switching to background mode would use the R = 6 pixel size and a smaller blue iris would be shown.

```
 1. measuring bkgnd
 2. At  775 347 Flux is:  11150.042862515344 CG [i,j] is: 775.09 347.0
 3. measuring star
 4. SNR=  292.5461666662765
 5. At  798 324 Flux is:  42041.093560102665 CG [i,j] is: 798.16 324.25
 6. R is now:  7
 7. measuring bkgnd
 8. At  774 347 Flux is:  8412.821864777645 CG [i,j] is: 774.19 347.0
 9. measuring star
10. SNR=  320.69763489441755
11. At  798 324 Flux is:  37827.66953405302 CG [i,j] is: 798.17 324.23
12. R is now:  6
13. measuring bkgnd
14. At  774 347 Flux is:  6355.388777080528 CG [i,j] is: 774.19 346.96
15. measuring star
16. SNR=  344.11234760870155
17. At  798 324 Flux is:  33788.263349415785 CG [i,j] is: 798.17 324.21
18. R is now:  5
19. measuring bkgnd
20. At  773 347 Flux is:  3947.0397439949434 CG [i,j] is: 773.13 346.9
21. measuring star
22. SNR=  356.4073564037814
23. At  799 324 Flux is:  26338.499600996074 CG [i,j] is: 798.8 324.16
24. R is now:  4
25. measuring bkgnd
26. At  773 346 Flux is:  2670.3271175726927 CG [i,j] is: 773.09 345.97
27. measuring star
28. SNR=  331.2372025077752
29.
```

FIGURE 10.4    Shows the console's output while adjusting the measurement iris. Reducing the iris to 5 pixels increased the SNR from 292 to 356, but the SNR dropped to 331 with a 4 pixel radius.

magnitudes 10.5 and 9.8, respectively, and SNR values of 331 and 366, respectively. The target's magnitude is estimated to be 9.83. The error is +/− 0.0 in this case because it is based on how good the linear fit is for the two reference stars, and since only two reference stars were used in this

```
 1. saving measurement #: 0
 2. +++   0 ['m', 10.5, '155858.27', '260804.6', 798.11, 324.08, 19787.07844081295,
2670.3271175726927, 331.2, []]
 3. measuring bkgnd
 4. At   624 765 Flux is:  2841.428571428572 CG [i,j] is: 623.98 764.93
 5. measuring star
 6. SNR=  366.0315013758386
 7. At   630 743 Flux is:  22352.756795195957 CG [i,j] is: 629.98 743.09
 8. saving measurement #: 1
 9. +++   1 ['K', 9.8, '155828.42', '255630.9', 629.98, 743.09, 22352.756795195957,
2841.428571428572, 366.0, []]
10. Doing calibration...
11. 10.5
12. 9.8
13. dmsq= 0.0
14. measuring bkgnd
15. At   944 644 Flux is:  2808.6995548164623 CG [i,j] is: 944.0 644.06
16. measuring star
17. SNR=  365.7793749365455
18. At   961 626 Flux is:  22193.969637866794 CG [i,j] is: 961.07 626.06
19. saving measurement #: 2
20. Target: Flux=  19385.27008305033 Magnitude =  9.83 +/- 0.0
21.
```

FIGURE 10.5    Console display for a two-reference star and one target measurement.

example for simplicity, it results in a perfect fit. Repeated measurements of the image magnitude with changes in the iris size and the manual centering cause the magnitude to vary by about +/− 0.05 mags, which isn't bad for a one-minute exposure of 10th magnitude objects with a jpg from a 50-mm smart telescope.

## CLASS IMPHOT PROGRAMMING NOTES

The **imphot** class is based on the **imcat** class previously described. It is more complicated, because it relies on interactive graphics, with settings adjustment – the aperture size – and is not automated; there is no simple thresholding to define a target or star, the user selects a target for measurement by clicking on it. Being interactive has the advantage that a background can be selected to avoid nearby stars which would cause the background level to be overestimated.

**imphot** is initialized as shown in Figure 10.6 (line 8), which has variables to track the star and background fluxes and areas, and flags to indicate if a circle should be drawn for either. The *_idx parameters are used to index into a list of values – mostly to demonstrate how this might be done and would be useful for later code enhancement and mostly unused for now. **C0** and **C1** are the linear coefficients returned by the regression analysis (linear fit) of the measured fluxes for the reference stars to their assumed magnitudes and once determined can be used to convert target fluxes into magnitudes.

Reference star data is read in from the star_mags.csv file into **self.star_list** using the **read_star_list()** function. Most of the fields are unused and

```
1. import matplotlib.pyplot as plt
2. import numpy as np
3. import csv
4. import math
5. import imcat as imc
6.
7. class imphot(imc.imcat):
8.     def __init__(self):
9.         super().__init__()
10.
11.         self.star_mags_file = ''
12.         self.R            = 8          # default aperture radius
13.         self.show_star_iris = False
14.         self.show_bgnd_iris = False
15.         self.iris_f        = 0         # star flux
16.         self.iris_b        = 0         # background flux
17.         self.star_iris_area = 0
18.         self.bgnd_iris_area = 0
19.         self.label_idx     = 0
20.         self.mag_idx       = 1
21.         self.hms_idx       = 2
22.         self.dms_idx       = 3
23.         self.icg_idx       = 4
24.         self.jcg_idx       = 5
25.         self.f_idx         = 6
26.         self.b_idx         = 7
27.         self.snr_idx       = 8
28.         self.mest_idx      = 9
29.         self.snr           = 0.0
30.         self.c0            = 0         # polynomial coeffs
31.         self.c1            = 0
32.         self.active_star   = 0
33.         self.mode          = ''
34.
35.
36.     def read_star_list(self):
37.         line_count = 0
38.         file_path = self.proj_dir  + self.star_mags_file
39.         with open(file_path, newline='') as csvfile:
40.             lines_in = csv.reader(csvfile, delimiter=' ', quotechar='|')
41.             for row in lines_in:
42.                 line_count += 1
43.                 [label,mag,hms,dms,icg,jcg,f,b,snr,mest] = row[0].split(',')
44.                 ra  = hms
45.                 dec = dms
46.                 # label, mag, ra, dec, icg, jcg,flux, bkgnd, m
47.                 self.star_list.append([label,float(mag),ra,dec,[],[],[],[],[],[]])
48.         self.Nstars = line_count
49.
50.     def draw_star_iris(self,r,i,j,circle_color):
51.         if self.show_star_iris == True:
52.             self.star_circle.remove()
53.         dc = plt.Circle( (j,i ), r, fill = False, color=circle_color )
54.         self.star_circle=self.ax.add_artist( dc )
55.         self.show_star_iris = True
56.
57.     def draw_bgnd_iris(self,r,i,j,circle_color):
58.         if self.show_bgnd_iris == True:
59.             self.bgnd_circle.remove()
60.         dc = plt.Circle( (j,i ), r, fill = False, color=circle_color )
61.         self.bgnd_circle=self.ax.add_artist( dc )
62.         self.show_bgnd_iris = True
63.
64.     def update_ref_star(self,i,j,icg,jcg):
65.         f = self.iris_f
66.         b = self.iris_b
67.         snr = (f-b)/np.sqrt(b)
68.         self.snr = snr
69.         self.star_list[self.active_star][self.icg_idx]=icg
70.         self.star_list[self.active_star][self.jcg_idx]=jcg
71.         self.star_list[self.active_star][self.f_idx]=self.iris_f
72.         self.star_list[self.active_star][self.b_idx]=self.iris_b
73.         self.star_list[self.active_star][self.snr_idx]=round(snr,1)
74.
```

FIGURE 10.6    Class **imphot**.                                        (*Continued*)

```
75.    def add_star(self,i,j,icg,jcg):
76.            f = self.iris_f
77.            b = self.iris_b
78.            m = self.show_magnitude(f-b)
79.            snr = (f-b)/np.sqrt(b)
80.            self.snr = snr
81.            star_data = [self.active_star,'na',
82.                         0, 0,
83.                         icg,jcg,
84.                         self.iris_f, self.iris_b, round(snr,1),
85.                         round(m,2)]
86.            self.star_list.append(star_data)
87.
88.    def measure_iris(self,event):
89.            i = 0; j = 0
90.            i = int(event.ydata)
91.            j = int(event.xdata)
92.            #print(i,j)
93.            icg = 0
94.            jcg = 0
95.            X = self.R
96.            Y = self.R
97.            R2 = X*X
98.            b = 0
99.            area = 0
100.           for x in range(-X, X):
101.               for y in range(-Y, Y):
102.                   r2 = x*x + y*y
103.                   if r2 < R2:
104.                       bpix = self.img[i+x, j + y]
105.                       b    += bpix
106.                       icg += bpix*(i+x)
107.                       jcg += bpix*(j+y)
108.                       area += 1
109.           icg = round(icg/b,2)
110.           jcg = round(jcg/b,2)
111.           self.iris_i = i
112.           self.iris_j = j
113.           self.iris_icg = icg
114.           self.iris_jcg = jcg
115.           return b,area
116.
117.   def save_measurement(self):
118.           i = self.iris_i
119.           j = self.iris_j
120.           icg = self.iris_icg
121.           jcg = self.iris_jcg
122.           if self.active_star < self.Nstars:
123.               self.update_ref_star(i,j,icg,jcg)
124.               print('+++ ',self.active_star, self.star_list[self.active_star])
125.               if self.active_star == (self.Nstars-1):
126.                   print("Doing calibration...")
127.                   self.calibrate()
128.           else:
129.               self.add_star(i,j,icg,jcg)
130.
131.
132.
133.   def calibrate(self):
134.           self.c0 = 0
135.           self.c1 = 1
136.           ivals = []                  # measured intensities
137.           mags  = []                  # known mags
138.           dmsq = 0
139.           for i in range(self.Nstars):
140.               [name,m,ra,dec,icg,jcg, s, b,snr,me] = self.star_list[i]
141.               d = s - b
142.               ivals = ivals + [d]
143.               mags = mags  + [m]
144.           self.c1, self.c0 = np.polyfit(np.log(ivals), mags, 1)
145.           for i in range(self.Nstars):
146.               m_est= self.get_magnitude(ivals[i])
147.               print(m_est)
148.               self.star_list[i][self.mest_idx] = m_est
149.               dmsq += (mags[i] - m_est)**2
150.           dmsq = dmsq / self.Nstars
```

FIGURE 10.6 (CONTINUED)   Class **imphot**.                              (*Continued*)

```
151.            print('dmsq=',dmsq)
152.            self.rms = np.sqrt(dmsq)
153.
154.     def show_magnitude(self, b):
155.            m = self.get_magnitude(b)
156.            print("Target: Flux= ", b, "Magnitude = ", round(m,2), '+/-',round(self.rms,3))
157.            return m
158.
159.     def get_magnitude(self, b):
160.            m = self.c0 + self.c1*np.log(b)
161.            return round(m,2)
162.
163.     def on_click(self,event):
164.            r = self.R
165.            if self.mode in ['star','bgnd']:
166.                f,area = self.measure_iris(event)
167.                i = self.iris_i
168.                j = self.iris_j
169.                if self.mode == 'star':
170.                    self.iris_f = f
171.                    self.star_iris_area = area
172.                    self.draw_star_iris(r,i,j,'white')
173.                    self.snr = (self.iris_f - self.iris_b)/math.sqrt(self.iris_b)
174.                    print('SNR= ',self.snr)
175.                elif self.mode == 'bgnd':
176.                    self.draw_bgnd_iris(r,i,j,'blue')
177.                    self.iris_b = f
178.                    self.bgnd_iris_area = area
179.                print('At ',i,j, 'Flux is: ', f, 'CG [i,j] is:' , self.iris_icg,self.iris_jcg)
180.
181.     def on_press(self,event):
182.            if (event.key == 's'):
183.                self.mode = 'star'
184.                print('measuring star')
185.            if event.key == 'b':
186.                self.show_iris = False
187.                self.mode = 'bgnd'
188.                print('measuring bkgnd')
189.            if event.key == 'n':
190.                self.show_iris = False
191.                self.mode = 'n'
192.                print("saving measurement #:", self.active_star)
193.                self.save_measurement()
194.                self.active_star += 1
195.            if event.key == '+':
196.                self.R += 1
197.                print ('R is now: ',self.R)
198.            elif event.key == '-':
199.                    self.R -= 1
200.                    print ('R is now: ',self.R)
201.
202.     def on_key_release(self, event):
203.            if event.key == 'shift':
204.                self.mode = ''
205.
```

FIGURE 10.6 (CONTINUED)    Class **imphot**.

are there to serve as a template to be modified if another reference star format source is used. The return value (**self.Nstars**) is the number of reference stars, that is, the number of stars that must be measured to set the magnitude calibration before other targets are studied.

There is no mechanism in this demonstration code to automatically select stars; the user must be able to identify the stars in the image and measure them in the order that they appear in the star_mags.csv file.

The **draw_star_iris()** and **draw_bgnd_iris()** functions allow for drawing and removing the white and blue iris circles, depending on whether the star or the background is being measured. The circles are added using the MATPLOT **add_artist** function with the return value stored in **self. star_circle** or **self.bgnd_circle**, which are then used for removing the last drawn blue or white circle when needed (see lines 50 and 57).

**self.star_list** is initially populated with reference stars whose entries are updated by the **update_ref_star** function as they are measured, and later, new entries added as targets are measured using the **add_star** function (see line 86).

Measuring a star consists of clicking on it, and the mouse coordinates (i,j) are used by the **measure_iris()** function (line 88) to analyze the iris centered on that location. Pixels in the iris are found by iteration over the square region of size +/− R centered on the mouse click, and within that square, only pixels within radius R of the center are used. The flux is estimated simply by adding all pixel fluxes, the area by counting pixels in the circle, and the center of gravity (icg, jcg) is calculated for a better estimate of the center than the user's mouse click. Measurements are saved to **self. star_list** using the **add_star** or **update_ref_star** previously described.

Calibrating magnitude is done by the **calibrate()** function (line 133) through applying a polynomial fit (set to a linear model in the code) to the results from measuring the reference stars. For each reference star, its starlight (the flux minus the background – see line 141) is regressed against the magnitude (line 144) yielding the coefficients for the fit: **C0** and **C1**. Magnitude has a logarithmic relationship to the star's brightness, and function **get_magnitude()** computes the magnitude for a specified brightness (line 159). The root mean square of the reference star residuals is reported, but since in our example only two reference stars were used, the fit will be perfect, and the error is zero. If more reference stars are provided, the error would be non-zero.

As a measure of image quality, the SNR is roughly estimated from the ratio of the star's amplitude above the background (Total flux – Background) divided by the square root of the background (see line 79). The best aperture can be determined by trying to maximize SNR prior to making a formal set f measurements.

Finally, keyboard and mouse events are supported with the **on_click**, **on_press**, and **on_key_release** routines that allow the user to switch between star and background measurement ('s' or 'b'), saving the measurement and moving on to the next star ('n'), and increasing or decreasing the iris size ('+ or −').

## OTHER CONSIDERATIONS

Photometry is very challenging to do properly, and it is easy to think your carefully made measurements must necessarily be reasonable and of high quality. The reality is, most likely, your first results will disappoint when you compare them with others. There are many possible reasons. A simple calculation of brightness or magnitude from fluxes that are compared to reference stars can be too simple. Optical systems and cameras can have different sensitivities to different wavelengths requiring complex calibration techniques. Perhaps one of the comparison star has an unknown variability? Perhaps your sensor is non-linear and some stars are saturating more than you realize? Perhaps neighboring light pollution is creating an unwanted gradient in your images? For reasons such as these, it is important to be realistic and honest in your accuracy estimates. One way to get a sense of your accuracy would be to measure the magnitudes for a set of stars within a few magnitudes of your target's. Then, their magnitudes' standard deviation is a reasonable estimate of your measurement error. My experiments using the **imphot** software on the 9th mag star T Crb (when it was anticipated to go nova in 2024) were consistent at the 0.05–0.07 mag level, using the Seestar 50 jpg images. This is a very reasonable result for a software solution that is very quick and easy to use, very transparent in its workings, and therefore very suitable for student use.

## SUMMARY

In the code we described here, we provided the user with a functioning interactive solution to learn from and to perhaps modify for their own purposes. Most photometry software solutions are very complex, with more capabilities than might be necessary, so there is a benefit in having a simpler version – even if not automated – to learn from, to work with. The beauty of organizing code into a python class structure is the user could build on the class, if they wanted to develop a custom solution, perhaps more automated, for their needs. Just as we did for the imcat class, we could add a simple function to save the measured data in self.star_list to a csv file if desired. We didn't do this step since it's not that difficult and is one less step to explain.

# Aligning Images and Finding Targets

O UR BRAINS ARE SUPERB pattern matching and image processing machines. For example, look at the two images of the T CrB region taken a day apart, shown in Figure 11.1. We can readily identify stars in one image corresponding to those in the other. It all seems so easy; the two images appear to be a simple shift or translation of each other. Surely there must be a simple algorithm to map one image onto the other.

Unfortunately, there might not be a simple solution in general. In our case, since the images were taken with an alt-az telescope (a Seestar 50), and not being polar aligned, different images of the same area can have significant field rotation, and we cannot assume none is present between these images; since these were taken around the same time on subsequent nights, at 10:00 pm and 9:55 pm, the field rotation is indeed small. If, however, the second image had been taken at 11 pm, a 15-degree rotation would exist between the two.

The problem we're facing in trying to match targets in one image with those in another is also compounded by the fact the images might not contain the same objects. Asteroids can be present in one, but not the other, or if in both images, they might have changed position. Variable stars and novae can be present on one and not the other, and stars might enter or move out of the field as the center changes. These factors ('transients') make it difficult to try something like using the center of mass of the star positions to determine a reference point and to see how it shifted

DOI: 10.1201/9781003600046-11

FIGURE 11.1    Two images of T Crb (slightly above the image centers) taken a day apart. Even though the images have shifted relative to each other, we can easily pick out matching star patterns.

during exposures – if the detected stars are not all the same, this approach will not be robust.

And there is yet another complication, if the images were taken using different telescopes and cameras, the plate scales and fields of view also need to be considered.

What these concerns reduce to is that while images could be aligned by eye, if one is made semi-transparent and it is moved and rotated over the other, and possibly stretched, it is a significantly more difficult problem to find a computer codable algorithm.

Our goal in this chapter is to explore how images might be aligned programmatically, without human intervention. Because we are trying to match detected objects in one image, with those in another, we will in fact be framing the problem as that of how we should match objects in one catalog, with those in another.

## IMAGE FEATURES

One approach to matching images is to identify features common to both and to use these to figure out how to map one image onto the other. While this seems simple, it's difficult to find a general solution that is robust, especially when the images might have significant differences.

Our brains are excellent at pattern matching and at detecting real or imagined patterns. What is it about a star in one image that makes it possible for our brains to identify it in another? If it does not have something intrinsic such as extreme brightness difference or color relative to others, then it must be the spatial relationships between the star and all others. This suggests exploring these relationships, which at the simplest, relates to the distances and directions between stars.

Appendix I shows a simple Python class (**spatial**) that graphically emphasizes the spatial relationships each star in an image has with its neighbors and a short code for running it. It is easily run and was applied to a T CrB image to produce the plots shown in Figure 11.2. Instead of simply drawing connecting lines between all possible stars, the plot shows the connections between each of the stars used and nearby neighbors (actually up to 6 were plotted to reduce clutter), by drawing short radials (vectors) where each radial length was scaled to be a third of the separation.

An important aspect of class **spatial** is that it uses a balanced tree solution (KDTree) which organizes the star positions efficiently for searching and can quickly identify the nearest neighbors to a specified position. The search was done within a specified radius – set to 300 pixels here after a little trial and error.

While the results are intriguing and suggest each star's radial patterns might serve as a 'fingerprint' to help identify it across multiple images, the plots are most useful in suggesting the existence of features based on vector separations that might serve our needs, in other words, we are looking for a solution that captures the spatial relationships in a *useful* form.

The patterns shown in Figure 11.2 are not easy to work with. For example, how should one compare one set of star radials with another, especially if they change because of the addition/subtraction of a radial because of a variable star or asteroid (radials are added/subtracted to a group)? How should one measure the angles presented by each radial? Perhaps relative to the longest one? How many neighbors should be included?

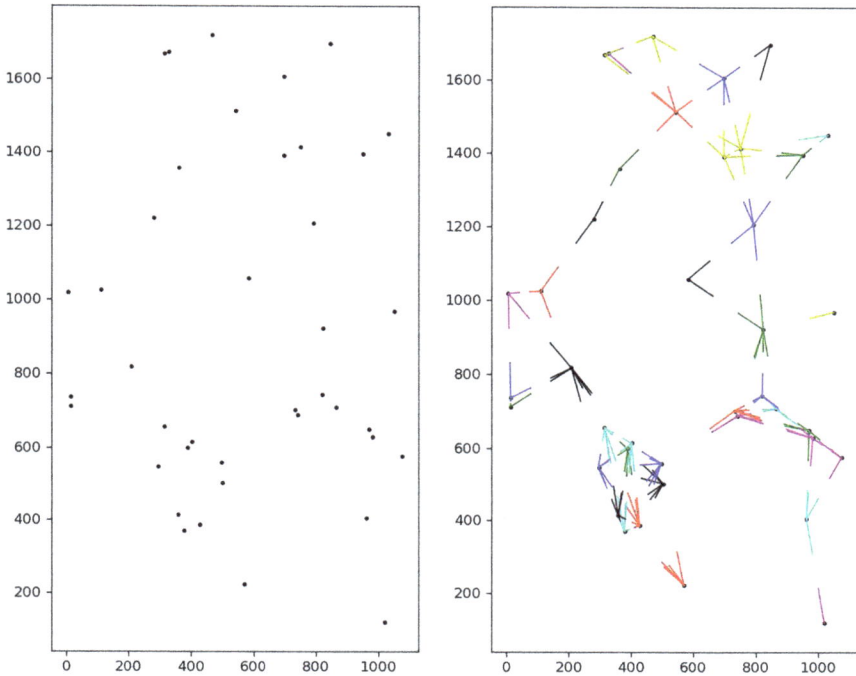

FIGURE 11.2 Using our class **spatial** allowed us to identify the neighbors for each star and draw radials which emphasize each star's relationship to its neighbors. The resulting patterns for each star suggest features based on separations and directions would be useful for identification purposes.

Fortunately, a solution to the general problem has been found based on the notion of constructing quadrilaterals. The general idea is that all possible combinations/groupings of four stars in an image are generated, and for each group (called a 'quad'), the two widest are used to define points (0,0) and (1,1) for the group. Then the other two points can have their coordinates expressed as (a,b) and (c,d). A quad is then numerically identifiable by a 4-digit group (a,b,c,d) and the numbers have magnitudes less than 1, so image magnification or telescope focal length effects are removed. In comparing two images, we only have to find quads in one that match quads in the other. Some quads will not match because of stars entering or leaving at the edges, or because of transients. But this is okay, because if there is sufficient overlap, those static targets in both images will create matching quads, and so, unlike the star radials we previously mentioned, using structures like quads is intrinsically more robust, since in creating our combinations, we will naturally be creating quads that do not include transients, and these will be available for matching.

Obviously, other geometric shapes could be used using other numbers of stars, instead of using 4; the principle would be the same, but it is not necessary for our purposes to use higher numbers, and we will demonstrate the process using triangles made from star triplets.

Our general strategy will be to:

1. Set the number of stars to be used from each catalog. This was set to 6 for our example. The number of combinations to test roughly scales as the cube of this number. Since the catalogs are sorted by brightness, the 6 brightest catalog entries were used from each catalog.

2. Read the brightest targets from the image catalogs into dataframes df1 and df2.

3. Create lists of three-star combinations for each image (combos1 and combos2), where a combination is a list like [sid1, sid2, sid3], that is, it consists of the star_id numbers in our catalog files.

4. For each combination from image 1, compare it to every combination for image 2, by

   a. Finding and sorting the lengths of the triangle sides and,

   b. Calculating the sum of the absolute values of the differences. For example, if the first triangle's sides are 2,3,4, and the second triangle's sides were 2.1, 3.2, and 3.9, the sum of the absolute differences would be (0.1 + .2 + .1) = 0.4.

5. Use the triangles with the smallest sum of differences as the best match and from these, the translational and rotational shifts can be determined.

In our approach, we are simply using the lengths of triangle sides for matching. Implicit in specifying the sides are the associated angles, so we really are relying on the star radials shown in Figure 11.2, but we are only using two radials at a time from any star.

The complete code for this solution (class **catalign**) is described below. We can see how well the **catalign** class functions from the small test program shown in Figure 11.3. It's very basic; two catalogs are imported and the align_ims function is used. After the alignment is finished, two

```
 1. from imcat_align import catalign
 2. import math
 3.
 4. CAT                = catalign()
 5. CAT.proj_dir       = './'
 6. N                  = 140            # number of stars to read from catalog
 7. Nstars             = 6              # number of brightest stars for matching
 8.
 9. CAT.fname1         = 'T_CrB_20240517'
10. CAT.df1            = CAT.read_csv_into_df(N,CAT.fname1)
11. df1                = CAT.df1
12.
13. CAT.fname2         = 'T_CrB_20240518'
14. CAT.df2            = CAT.read_csv_into_df(N,CAT.fname2)
15. df2                = CAT.df2
16.
17. df2s,df2sr,tri1,tri2 = CAT.align_ims(Nstars,df1,df2)
18.
19. print ("Lin. Offset (pix) = ",round(CAT.di,4), round(CAT.dj,4))
20. print ("Rot. Cent.  (pix) = ",round(CAT.Icen,4), round(CAT.Jcen,4))
21. print ("Rot. Offset (deg) = ",round(CAT.theta_rad*180/math.pi,4))
22.
23.
24. ###############  Compare original coordinates ############################
25.
26. text='Original catalogs: im1 (red) and im2 (blue)'
27. CAT.plot_AB_coordinates(df1,df2,text,'./originals.jpg')
28.
29. ######################### Plot field stars and matching triangles #########
30.
31. CAT.plot_triangles(df1,tri1,df2,tri2,'./matching triangles.jpg')
32.
33. ######################### plot im1 and shifted im2 ####################
34.
35. text='Linear shift of im2'
36. CAT.plot_AB_coordinates(df1,df2s,text,'./shifted.jpg')
37.
38. ######################### plot im1 and shift+rotated im2 stars #########
39.
40. text='Shift and Rotation: im1 (red) and im2 (blue)'
41. CAT.plot_AB_coordinates(df1,df2sr, text,'./shifted_and_rotated.jpg')
42.
```

FIGURE 11.3 A short program **test_imcat_align.py** to compare and align catalogs for two images. New dataframes are created with shifted (df2s) and shift and rotated (df2sr) versions of df2. The linear and rotation offsets are printed to the console.

additional dataframes have been created and are available for further study or plotting. The reference catalog is stored in **df1**, and the one being matched is in **df2**. After the alignment, the linearly shifted version of **df2** is saved as **df2s**, and the rotationally corrected version of **df2s** is saved as **df2sr**.

Controlling parameters were the number of stars to read in from each catalog (140 in this example), and how many of the brightest stars to use for matching (we used 6 here). The simplicity of the code suggests it would be easy to modify, so a folder or directory of catalogs could be matched.

To visually see how the alignment process was working, plot instructions were included at the end of the test program. These were used to create four plots: 1. A plot showing the two raw catalog entries overlapped (line 27); 2. A plot showing the best fit matched triangles (line 31); 3., A plot showing the linearly shifted **df2** (now called **df2s**) over the **df1** data

Original catalogs: im1 (red) and im2 (blue)

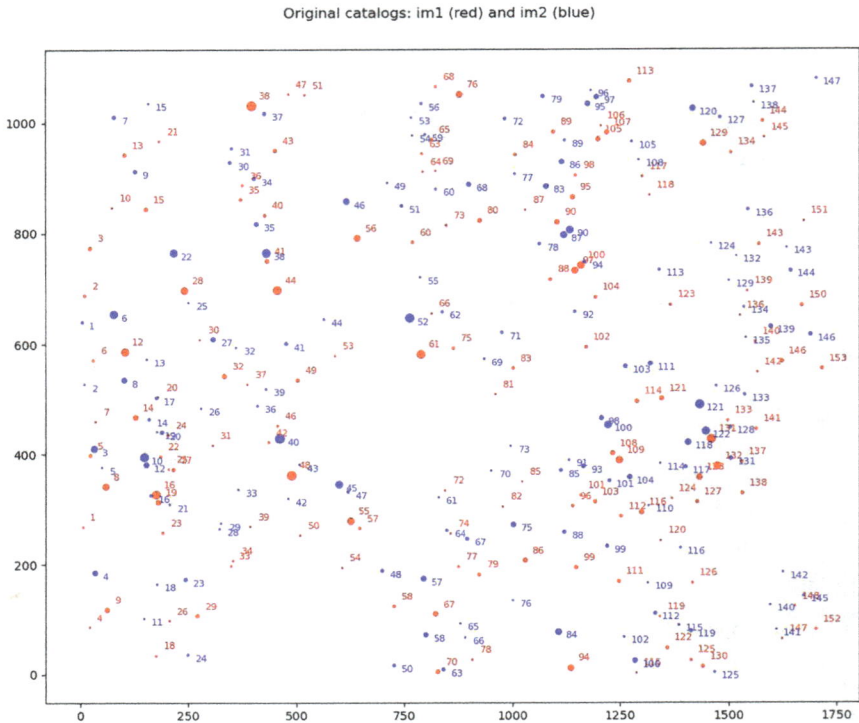

**FIGURE 11.4** Here the objects from df1 (red) and df2 (blue) are plotted so their relative shift can be seen. T CrB is the red/blue pair labelled 61/52 near the center.

(line 36), and 4. An overlay of the shifted/rotated **df2sr** data over the df1 data to show the overall results (line 41).

The raw plots are shown in Figure 11.4, where the blue stars are the **df2** objects and the red stars the **df1**. T CrB is the red/blue 61/52 pair near the center, which clearly shows there is a definite shift between the images. (Plotted objects show color coded labels to show which catalog was used for each label).

In Figure 11.5, the star triplets used to match the catalogs are shown. The linear shift between the images is found by subtracting the coordinates of sid 61 (red) from those of sid 52 (blue). The angle offset for a triangle side is found by comparing its directions for the two catalogs, that is, the directions from sid 10 to sid 52 with that for sid 16 to sid 61. The rotational offset is the average from the three sides.

In Figure 11.6, a zoomed in portions of the shifted **df2s** data and the **df1** plots is shown. It just so happened that T CrB was used for the linear shift (it was one of the six brightest objects in both catalogs), and as the result

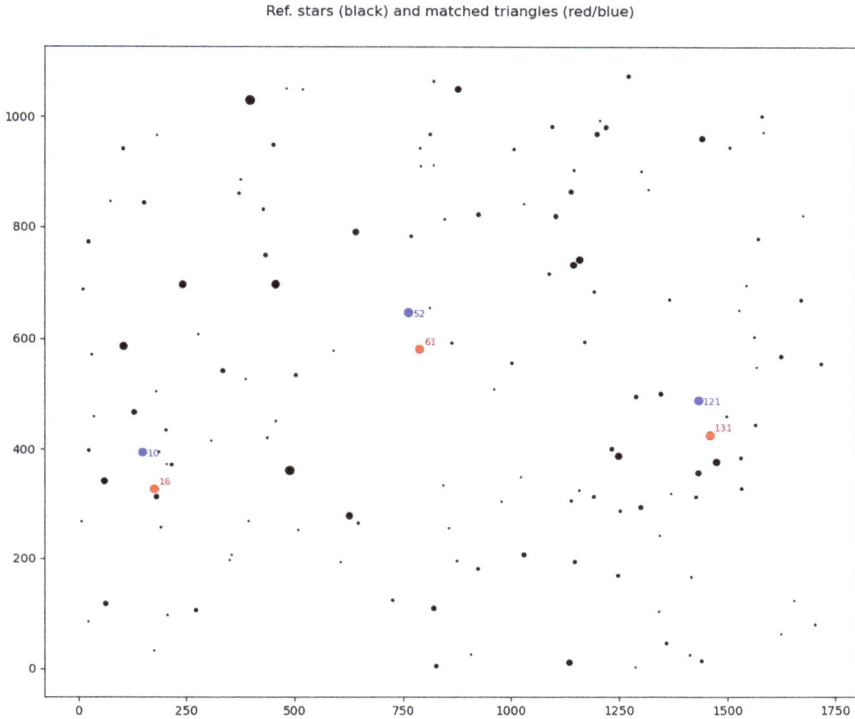

Ref. stars (black) and matched triangles (red/blue)

FIGURE 11.5 The triangle/triplet stars from the df1 (red) and df2 (blue) catalogs used for alignment.

shows excellent alignment, but for nearby stars, there is an uncorrected rotation, and they show up as offset red/blue dot pairs.

Finally, adding in the rotation correction fixes the unaligned stars (see Figure 11.7), which shows a zoomed-in portion of the region near the bottom right. The uncorrected and corrected versions for the rotation are shown on the left and right panels.

Important outputs from our alignment efforts are the estimates of the linear and rotation offsets, and the center of rotation. Knowing these, equations could be written to map the coordinates from one catalog into the other's reference frame. This would allow for a sequence of images to be analyzed where a target of interest was identified in the reference (first) image, and then knowing the transformation equations, analyze the corresponding region in each of the subsequent images. The analysis might be photometric, such as measuring the intensity centered on the calculated position.

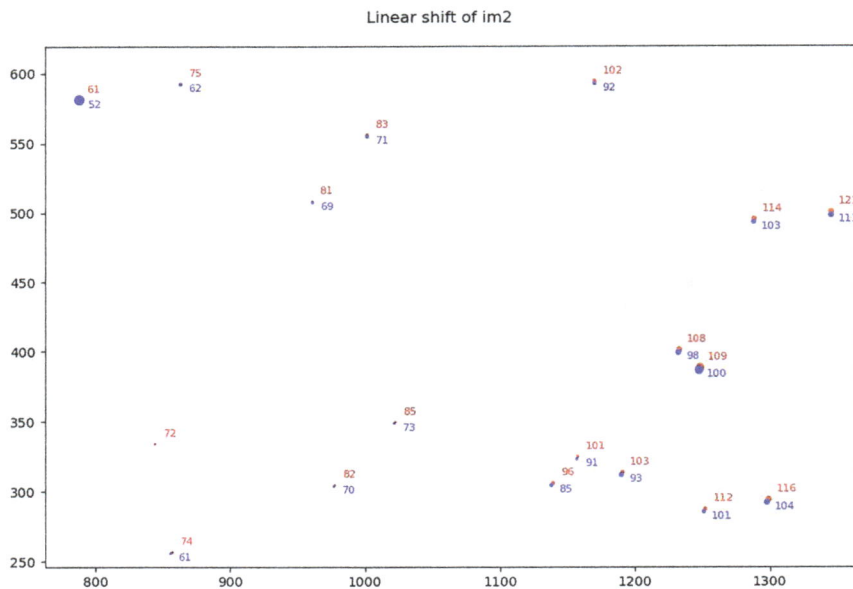

FIGURE 11.6   Close up of the region near T CrB (object 61/52 on the lower left). After the linear shift, which happened to use T CrB as the center. T CrB aligned just fine, but other stars show an uncorrected rotational offset.

But there is another approach to using our alignment results; instead of studying the comparison images based on pixel coordinates values derived from the transformation equations, use the alignment to identify catalog entries to match targets. There is an important difference between these two approaches. In the first, the target's properties are measured from an estimated position based on the transformation equations; in the second, by using transformation equations to identify the target in the comparison image, the target's cataloged/measured position can be used for the analysis, and the catalogued position is probably the more accurate.

For our demonstration we compared two catalogs derived from our images. There is no reason why one of the catalogs could not be a subset of a standard star catalog, suitably modified, using astronomical equatorial coordinates (Right Ascension and Declination). This way target equatorial coordinates could be found and associated with our catalog targets, and equatorial coordinates added to all our catalog entries. Once done, images and catalogs could be accessed based on equatorial coordinates.

Finally, when constructing an overlay chart for two aligned catalogs, like the properly aligned chart (right pane) in Figure 11.7, each target has two labels, a red and blue, showing the sids/labels from their respective

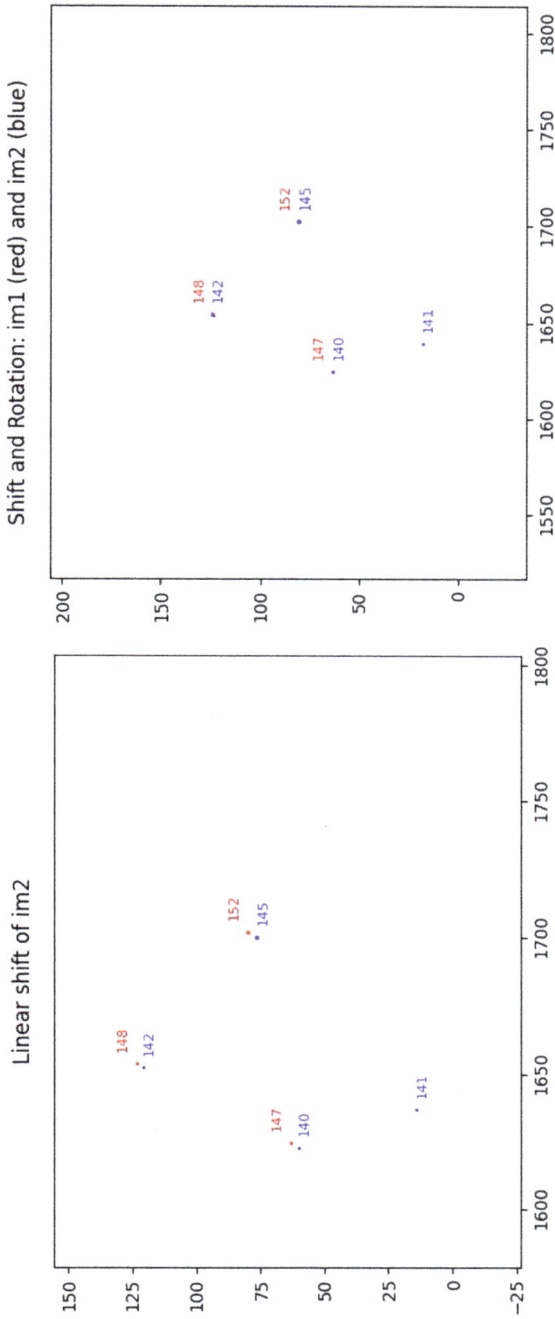

FIGURE 11.7  A close inspection of the region's bottom right corner shows the effects of adding the rotation correction (right side) to the linearly shifted catalog (left).

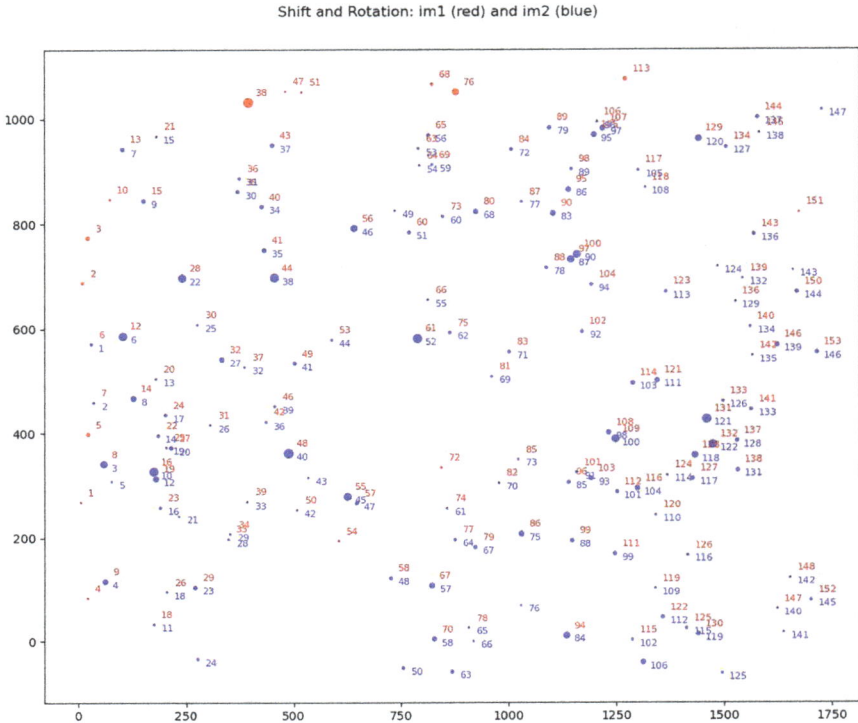

FIGURE 11.8 When properly aligned, the catalog objects present as single stars/targets with two labels. There are a few single label objects near the center, just below T CrB (object 61/52), for example the red one labelled 72. Without further study, being small, it's likely simply a marginal detection and only appearing in the first catalog, but it could be a transient.

catalogs. Any target with only one label is a transient or a moving target. For example, Figure 11.8 shows the zoomed-out overlay of the aligned catalogs. There are single label objects along the top and bottom showing the top objects were not present in the second catalog, probably because of image center shift. However, if not near the frame edge, single label objects warrant further inspection since that would be indicative of a transient object appearing in one catalog and not the other and therefore have only one label. At fainter levels, the transient might simply be a consequence of thresholding level, or transparency, but a brighter one could be real. For visual detection of transients with varying brightness, it might be useful to draw circles, instead of disks, since a change in brightness would result in two concentric circles of different sizes being displayed instead of two same-sized ones.

## CLASS CATALIGN PROGRAMMING NOTES

Class **catalign** (shown in Figure 11.9) is designed to compare two catalogs containing lists of detected targets from two images. Its main purpose is to determine the linear and rotational offsets between images taken with the same optical system (i.e., at the same plate scale) although, it could be modified to work with images taken with different optical systems.

```python
 1. import pandas as pd
 2. import math
 3. import numpy as np
 4. import matplotlib.pyplot as plt
 5. import itertools
 6.
 7. class catalign:
 8.     def __init__(self):
 9.         self.proj_dir      = ''
10.         self.fname1        = ''
11.         self.img_file1     = self.fname1 + '.jpg'
12.         self.fname2        = ''
13.         self.img_file2     = self.fname2 + '.jpg'
14.         self.df1           = pd.DataFrame()
15.         self.df2           = pd.DataFrame()
16.
17.     def read_csv_into_df(self,N,fname):
18.         fpath = self.proj_dir  + fname +'_cat.csv'
19.         df    = pd.read_csv(fpath, nrows=N)
20.         df.columns = ["sid","T","i","j","Flux","SNR"]
21.         return df
22.
23.     def align_ims(self,Nstars,dfA,dfB):
24.         tri1, tri2       = self.find_best_match_triangle(dfA,dfB,Nstars)
25.         self.get_lin_shift(tri1[0], tri2[0],dfA,dfB)
26.         df2s             = self.apply_lin_shift_BtoA(self.di,self.dj,dfA,dfB)
27.         self.theta_rad   = self.get_rot_offset(dfA,tri1,dfB,tri2)
28.         (Icen,Jcen,F)    = self.get_sid_ijF(tri2[0],df2s)
29.         df2sr            = self.apply_rot_offset(-self.theta_rad,Icen,Jcen,df2s)
30.         self.Icen = Icen
31.         self.Jcen = Jcen
32.         return df2s,df2sr,tri1,tri2
33.
34.     def find_best_match_triangle(self,df1,df2, Nstars):
35.         sids1     = df1['sid'][0:Nstars]
36.         sids2     = df2['sid'][0:Nstars]
37.         combos1   = list(itertools.combinations(sids1,3))
38.         combos2   = list(itertools.combinations(sids2,3))
39.         diff_list = []
40.         triplets  = []
41.         for i,c1 in enumerate(combos1):
42.             (s1a,s1b,s1c)    = c1
43.             sides1 = self.get_triangle_sides_from_sids(df1,s1a,s1b,s1c)
44.             for j,c2 in enumerate(combos2):
45.                 (s2a,s2b,s2c) = c2
46.                 sides2    = self.get_triangle_sides_from_sids(df2,s2a,s2b,s2c)
47.                 diff      = sum(abs(np.array(sides1) - np.array(sides2)))
48.                 diff_list.append(diff)
49.                 triplets.append([i,j])
50.         matching_combos = triplets[diff_list.index(min(diff_list))]
51.         #
52.         # triangles (triplets) of sids in im1 matching triplet in im2
53.         #
54.         tri1         = combos1[matching_combos[0]]
55.         tri2         = combos2[matching_combos[1]]
56.         return list(tri1), list(tri2)
57.
58.
59.     def build_df_using_new_ij_vals(self,df,ir,jr):
60.         dfnew = df.copy(deep=True)
61.         del dfnew['i']
62.         del dfnew['j']
```

FIGURE 11.9    Class **catalign**.                           (*Continued*)

```
63.            dfnew['i'] = ir
64.            dfnew['j'] = jr
65.            dfnew = dfnew[['sid','T','i','j','Flux','SNR']]
66.            return dfnew
67.
68.    def get_ij_sep(self,i1,j1,i2,j2):
69.            di = i1 - i2
70.            dj = j1 - j2
71.            d = math.sqrt(di*di +dj*dj)
72.            return d
73.
74.    def get_sid_ijF(self,sid, df):
75.            d0 = df[df['sid'] == sid]
76.            rlist = d0.values.tolist()
77.            [[sid2,T,i,j,Flux,SNR]] = rlist
78.            return [i,j,Flux]
79.
80.    def get_ijsF_df_cols(self,df):
81.            v1      = df['i'].tolist()
82.            v2      = df['j'].tolist()
83.            v3      = df['sid'].tolist()
84.            v4      = df['Flux'].tolist()
85.            return v1,v2,v3,v4
86.
87.    def get_triangle_sides_from_sids(self,df,s1,s2,s3):
88.            [i1,j1, F1] = self.get_sid_ijF(s1,df)
89.            [i2,j2, F2] = self.get_sid_ijF(s2,df)
90.            [i3,j3, F3] = self.get_sid_ijF(s3,df)
91.            d12     = self.get_ij_sep(i1,j1,i2,j2)
92.            d13     = self.get_ij_sep(i1,j1,i3,j3)
93.            d23     = self.get_ij_sep(i2,j2,i3,j3)
94.            dmin    = min(d12,d23,d13)
95.            sides   = [d12/dmin,d23/dmin,d13/dmin]
96.            sides.sort()
97.            return sides
98.
99.    def get_lin_shift(self,sid1,sid2,df1,df2):
100.            [i1,j1, F1] = self.get_sid_ijF(sid1,df1)
101.            [i2,j2, F2] = self.get_sid_ijF(sid2,df2)
102.            self.di     = i2 - i1
103.            self.dj     = j2 - j1
104.
105.    def apply_lin_shift_BtoA(self,di, dj, dfA,dfB):
106.            [i2,j2,sid2,f0] = self.get_ijsF_df_cols(dfB)
107.            x2s             = np.array(i2) - di
108.            y2s             = np.array(j2) - dj
109.            dfBs            = self.build_df_using_new_ij_vals(dfB,x2s,y2s)
110.            return dfBs
111.
112.    #
113.    # Get the orientations of the triangle sides
114.    # Use these to compare triangle orientations among images
115.    #
116.    def get_triangle_side_angles(self,sidlist,df):
117.            [s1,s2,s3] = sidlist
118.            [i1,j1,F1]      = self.get_sid_ijF(s1,df)
119.            [i2,j2,F2]      = self.get_sid_ijF(s2,df)
120.            [i3,j3,F3]      = self.get_sid_ijF(s3,df)
121.            ang21   = math.atan2(j2 - j1, i2-i1)
122.            ang31   = math.atan2(j3 - j1, i3-i1)
123.            ang32   = math.atan2(j3 - j2, i3-i2)
124.            return ang21,ang31,ang32
125.
126.    def get_rot_offset(self,df1,tri1,df2,tri2):
127.            a,b,c       = self.get_triangle_side_angles(list(tri1),df1)
128.            A,B,C       = self.get_triangle_side_angles(list(tri2),df2)
129.            avg_offset = (A-a + B-b + C-c)/3.
130.            return avg_offset
131.
132.    def apply_rot_offset(self,angle,Icen,Jcen,df):
133.            [i2,j2,s2,f2] = self.get_ijsF_df_cols(df)
134.            ir,jr           = self.rot_ij(Icen,Jcen,np.array(i2),np.array(j2),angle)
135.            df2sr           = self.build_df_using_new_ij_vals(df,ir,jr)
136.            return df2sr
137.
```

FIGURE 11.9 (CONTINUED)   Class **catalign**.                    (*Continued*)

```
138.    # Rotate arrays i1,j1 by angle centered on (icen,jcen)
139.    def rot_ij(self,icen,jcen,i1,j1,angle):
140.        c = math.cos(angle)
141.        s = math.sin(angle)
142.        i = c*(i1-icen) - s*(j1-jcen)
143.        j = s*(i1-icen) + c*(j1-jcen)
144.        i = i + icen
145.        j = j + jcen
146.        return (i,j)
147.
148.
149.    def plot_coordinates(self,df):
150.        fig, ax = plt.subplots()
151.        x1,y1,sid1,f1 = self.get_ijsF_df_cols(df)
152.        s1            = [int(f/1000)+1 for f in f1]
153.        ax.scatter(x1, y1, color = 'red', s=s1)
154.        for i, txt in enumerate(sid1):
155.            ax.annotate(txt, (x1[i], y1[i]), color='red',fontsize=8,\
156.                        xytext = (8,8),textcoords="offset pixels")
157.
158.    def plot_sidlist(self,sidlist,df,offset,color):
159.        x0 = []; y0 = []; s0 = []
160.        for s in sidlist:
161.            [i0,j0,F0]=self.get_sid_ijF(s,df)
162.            x0.append(i0)
163.            y0.append(j0)
164.            s0.append(int(F0/1000)+1)
165.        sizes = np.array(s0)
166.        plt.scatter(x0,y0,marker='o', color=color,s=sizes)
167.        for i,txt in enumerate(sidlist):
168.            plt.annotate(txt, (x0[i], y0[i]), color=color,fontsize=8,\
169.                        xytext = offset,textcoords="offset pixels")
170.
171.    def plot_AB_coordinates(self,dfA,dfB,txt, img_filename):
172.        mydpi=100
173.        fig=plt.figure(figsize=(1200/mydpi,1000/mydpi),dpi=mydpi)
174.        fig.suptitle(txt)
175.        x1,y1,sid1,f1 = self.get_ijsF_df_cols(dfA)
176.        x2,y2,sid2,f2 = self.get_ijsF_df_cols(dfB)
177.        s1            = [int(f/1000)+1 for f in f1]
178.        s2            = [int(f/1000)+1 for f in f2]
179.        plt.scatter(x1, y1, color = 'red', s=s1)
180.        for i, txt in enumerate(sid1):
181.            plt.annotate(txt, (x1[i], y1[i]), color='red',fontsize=8,\
182.                        xytext = (8,8),textcoords="offset pixels")
183.        plt.scatter(x2, y2, color='blue',s = s2)
184.        for i, txt in enumerate(sid2):
185.            plt.annotate(txt, (x2[i], y2[i]), color = 'blue',fontsize=8,\
186.                        xytext = (8,-8),textcoords="offset pixels")
187.        plt.savefig(img_filename,dpi=mydpi)
188.
189.    def plot_triangles(self,dfA,triA,dfB,triB, img_filename):
190.        [i1,j1,s,f1]    = self.get_ijsF_df_cols(dfA)
191.        s0              = dfA['Flux'].tolist()
192.        s1              = [int(s/1000)+1 for s in s0]
193.
194.        mydpi=100
195.        fig=plt.figure(figsize=(1200/mydpi,1000/mydpi),dpi=mydpi)
196.        fig.suptitle('Ref. stars (black) and matched triangles (red/blue)')
197.        plt.scatter(i1, j1, marker='o', color='black', s=s1)   # draw field stars
198.        self.plot_sidlist(triA,dfA,(5, 5),'red')
199.        self.plot_sidlist(triB,dfB,(5,-5),'blue')
200.        plt.savefig(img_filename,dpi=mydpi)
201.
```

FIGURE 11.9 (CONTINUED)   Class **catalign**.

In general, normal matrix i-j coordinates are used instead of cartesian x-y coordinates which places the coordinate origin at the top left instead of the bottom left corner. This means the plots will be rotated. In other chapters, we avoided this by converting **(i,j)** to **(x,y)** using a mapping like: $x = j$, $y = Nrows - i$. This slight overhead is not needed, nor used here, for simplicity.

Naming conventions for dataframes follow the processing flow. The reference image is **dfA** or **df1**, and the one being matched is **dfB** or **df2**; applying a linear shift to **df2** produces a new dataframe named **df2s**; adding a rotation to **df2s** produces **df2sr**.

Catalog and dataframe rows contain a label called 'sid' – short for **star_id**. Generally, stars are referenced using their **sid** instead of their dataframe row.

The class breaks down into five functional groups: Initialization and Input; top level functions to implement the alignment and find matching star triplets; utility functions for manipulating/selecting pixels based on their (i,j) coordinates; finding and applying linear shift and rotation corrections; and output plotting.

When reading in a catalog from a csv file, only the first N rows are read. With different catalogs, subsets could be selected after reading them in, and then sorting by brightness or size perhaps.

Function **align_ims** compares the dataframes for the catalogs. It is important to specify how many stars to use for matching – which is not the same as the numbers of stars to read in from the catalogs – since these are the stars from which the triplet combinations are created, and cross-compared to find the best match. Because the number of combinations scales as the third-power of the number to use for matching, only a small number should be used. **align_ims** finds the best matching triangles and uses them to get and apply the linear and rotation corrections.

Finding the best match is done by the **find_best_match_triangle** function. It uses the **itertool** library to build combinations (triplets) of **sids** for each catalog and then tests them against each other to find the triangles where the sum of the sides best agrees. When finished, it returns the triangles to be used for determining the liner and rotational offsets.

When **align_ims()** finishes, it returns dataframes for the shifted and rotated comparison image, and the star triangles used to make the mappings.

Finding the best matching triplet consists of testing all combinations for image 1 against those for image 2, and a list (**triplet**) stores each triangle pair (one from each catalog), at line 50, and also a list (**diff_list**) of the calculated differences (line 51). The best match triangles are found by using the index of the smallest difference to locate the corresponding triangle pair in the **triplets** list.

To support data manipulation and selection, there is a group of utility functions needed to build dataframes derived from others, and for extracting values from dataframes based on **sid** (**get_sid_ijF**) or data column labels (**get_ijsF_df_cols**), and to get the distance between points (**get_ij_sep**). Because processing creates new dataframes based on manipulating previous ones, deep copies are needed to protect data integrity and ensure the derived dataframes are independent of the originals and not just links.

Once a matching pair of triangles is found, they are used to derive the linear and rotational shifts needed to bring the catalogs into alignment. **get_triangle_sides_from_sids** takes the **sid**s that make up a triangle and returns a list of the triangle side lengths, normalized so the smallest is 1. The normalization is included in case catalogs if images of differing plate-scale were being compared and not strictly needed here.

Note, when comparing images with different plate-scales, the scale factor could be reasonably estimated from the ratio of the actual lengths (borders) of the triangles, and then used later as an additional correction (i.e., knowing the plate scale would be needed for creating plots showing targets from both catalogs), but catalog matching does not require it.

The linear (translation) offset between the images is found (**get_lin_shift**) by subtracting the coordinates of the first point of the matching triangles from each other (lines 102–103) and applied (**apply_lin_shift_BtoA**) to **df2** to create **df2s** (line 26). And similarly for rotations where, since the triangles now have one overlapping point, the orientation of a side can be compared between the triangles to find their directional discrepancy, which is used to map **df2s** onto **df2sr** by applying a rotation correction (line 29).

Finally, for demonstration and testing, there is a group of plotting functions. Function **plot_coordinates** will plot a dataframe's star positions, **plot_sid_list** will plot a list of **sid**s, and **plot_AB_coordinates** will plot the coordinates from two dataframes on the one chart. For convenience, the **plot_triangles** will add the points used by the matching triangles. The actual size of the drawn star is based on its brightness, with a minimum of 1.

## SUMMARY

Image alignment ('plate solving') has become an essential feature offered by most modern astrophotography systems, and is essential for smart-scopes that highly automate and simplify telescope imaging. It was long recognized as an extremely difficult problem and only solved in recent decades through innovative pattern matching techniques that had to be robust, insensitive, to scale and rotation changes, exposure and color filter differences, image center offset differences and transients causing, most of which meant the task of comparing targets in different often involved image catalog collection with different entries.

In this chapter, we demonstrated how a slightly simpler implementation of the quad-based method worked and found it could match up target in one image with another. Our code could be used to follow a target at a fixed relative position, such as a variable star or nova, across a sequence of images in order to follow its light curve. On the other hand, unmatched targets could be explored to see if they were transients or moving (asteroids/comets). In either case, it is important for students who might rely on such software to have some appreciation of how it works, and there are many possible experiments possible where other modification to quad technique could be explored and additional scripts to support automating image collections developed.

## CHAPTER 12

# The Saha Equation and the Balmer Spectrum

O<small>NE OF THE MOST</small> wonderful discoveries in astronomy was the realization a star's spectrum (a display of what colors or wavelengths it emits or absorbs, see Figure 12.1) contained information about the star's composition – at least at the surface (visible) layers. Students now learn that spectra can be either continuous (where the intensities change relatively slowly with wavelength) or can contain discrete features (spectral lines) or be a mix. Spectral lines indicate the presence or absence of particular colors. When present, there is a bright line in the spectrum; when absent, there is a black line visible against a bright background. The patterns these lines make is determined by which atoms are present, and the temperature, since every atom emits or absorbs a unique set of colors, controls the degree to which they are in play.

What makes the lines unique to an atom? The answer is atoms have positively charged protons in their nuclei that are surrounded by negatively charged electrons that occupy various electron levels. Atoms of different elements have different numbers of protons, and the complex interaction between the many charged particles will cause available electron-level energies to vary from one element to the next. But electrons can only orbit in particular levels; they can move between levels by emitting or absorbing particular energies, which is why an atom can emit or absorb (block) particular wavelengths of light. Temperature plays a role because it determines how fast atoms are moving and how hard they bump

DOI: 10.1201/9781003600046-12

FIGURE 12.1    Hydrogen, Balmer, emission spectrum.

into each other, causing electrons to jump to higher levels or break free altogether. Normally, electrons will fall back to a lower level if they can, within millionths of a second, emit a photon of light in the process with an energy matching the energy loss in moving to a lower energy state, resulting in emission spectra.

So, there can be very complex interactions at play; collisions might cause electrons to move to higher levels or even break free. Electrons can fall from higher levels into many successive lower ones; free electrons can be captured into a level and then cascade down through lower ones to get to the lowest available one. And atoms can become ionized, where electrons are stripped away.

This means that not only does the structure of an atom influence the photon energy transactions, but the amount of transactions between various levels is dictated by the temperature which controls which levels are active. At extremely cold temperatures, no visible light might be emitted at all since atoms can be moving so slowly, collisions are too weak to bump electrons to higher levels.

Our goal in this chapter is to model the simplest and most numerous element in the universe, hydrogen, which has one proton and one electron, and to show how its spectral lines change intensity with temperature, as the roles of different energy states change with changing temperature.

## MODELLING SPECTRAL LINES

Light emissions from a gas of hydrogen atoms will depend on its temperature, since temperature, ultimately, is simply a measure of how fast atoms and molecules move – this is why there is an absolute zero temperature (–273 degrees Celsius) since once particles stop moving, they can't move any slower. The hotter a gas, the faster the particles move, and the harder the impacts between atoms that drive electrons to higher levels (and even free) where they can fall back to lower ones and release photons of different wavelengths and energies, that show up as an emission spectrum such as that shown in Figure 12.1. (Note the spectrum in Figure 12.1 is only the pattern of lines in the visible part of the spectrum; with other temperatures, other wavelength patterns can exist, outside what we can see with the naked eye.)

To model these complex interactions for the hydrogen atom, we need to be able to link the atom's structure which dictates the possible wavelengths that can be emitted/absorbed, to the actual distribution of electrons among possible levels, since the numbers moving between levels generate the actual emissions. However, the number of atoms that can participate in emitting light can change with temperature – some fraction will have their electrons totally stripped; at very high temperatures, most hydrogen atoms will be fully ionized. (Note: It is beyond the scope of this book to derive the equations needed for this modelling so we will simply present them and show how they work.) The Balmer spectrum is seen when electrons fall from higher levels down to the second level, and a Lyman spectrum when they fall down into the first (ground) level.

The Saha equation tells us the ratio between the numbers of atoms in consecutive ionization states. For multi-electron atoms, there can be many possible states as electrons are stripped one by one from the atom. For hydrogen, only two states are possible: Neutral, and fully ionized, with the numbers in each being referred to as $N_I$ and $N_{II}$.

The Saha equation, using the atom's ionization energy ($\chi$, the energy needed to strip the electron from the atom), gives us a relationship between the different ionization states, using typical fundamental constants (e.g., mass of the electron $m_e$, Boltzmann constant k, and Plank's constant h)

$$\frac{N_{i+1}}{N_i} = \frac{2kTZ_{i+1}}{P_e Z_i} \left( \frac{2\pi m_e kT}{h^2} \right)^{3/2} e^{-\frac{\chi}{kT}} \tag{12.1}$$

which can be reduced to:

$$\frac{N_{i+1}}{N_i} = \frac{C}{P_e} T^{2.5} e^{-\frac{\chi}{kT}} \tag{12.2a}$$

Or, more specifically for the hydrogen atom,

$$\frac{N_{i+1}}{N_i} = \frac{0.033298}{P_e} T^{2.5} e^{\frac{-157878}{T}} \tag{12.2.b}$$

Here C is a constant made from combining together the numerous ones outside the exponents in Equation 12.1 and is 0.033298. In Equation 12.2b, we used $\chi/k = 13.6eV/k = 157878$. We have assumed $Z_{i+1}/Z_i$ is 1/2 for hydrogen (assuming $Z_2$ is 1 for a free proton and $Z_1$ is 2 for a bound electron) and $P_e$ is the electron gas pressure (typically between 0.1 and 100). For the hydrogen atom, we could write the left side of the equation as $N_{II}/N_I$ since only the un-ionized and ionized states are possibilities.

We'd like to know the fraction of atoms that are not ionized (since these are the only ones that create the Balmer or Lyman emission lines), which is given by:

$$f = \frac{N_I}{N_{tot}} = \frac{N_I}{N_I + N_{II}} = \frac{1}{1 + \dfrac{N_{II}}{N_I}} \qquad (12.3)$$

Using Equation 12.2, with 13.6 eV for hydrogen's ionization energy, then for a given temperature, and assumed $P_e$, we can now calculate $N_{II}/N_I$ and then f from Equation 12.3.

We will use a notation where we will refer to the number of atoms in a particular state k (i.e., with electrons at level k) as $n_k$. An emitted spectral line results from electrons falling to a lower energy level and releasing the energy difference in the form of a photon with that same energy, and that energy will determine the photon's wavelength and frequency. Because levels are unevenly spaced where energy is concerned, the transitions between them will involve unique energies and wavelengths. For example, the intensity of a line for a transition from level 5 into level 2 will depend on how many atoms are in the level 5 energy state and so we would need to know what $n_5$ was.

We still need to know the atomic energy states, so we can estimate intensities. Boltzmann statistics show the relative numbers of atoms at energy states 'b' compared to energy states 'a' (see Equation 12.4). It depends on the temperature T and the energies of the individual states. The terms $g_a$ and $g_b$ are simply the capacities (degeneracies) of the states – how many electrons can exist at a given state. For hydrogen, g takes on the values $2n^2$ (n = 1, 2, 3…) so its levels, starting from the lowest, hold 2, 8, 18 electrons and so on. Note, if T is very small, the exponent term goes to zero, while if T becomes very large, the exponent term becomes 1.

Equation 12.4 gives a ratio of populations needed by our models since we will assume the intensity is proportional to the population size.

$$B(a,b) = \frac{n_b}{n_a} = \frac{g_b}{g_a} e^{\frac{-(E_b - E_a)}{kT}} \qquad (12.4)$$

If we assume most of the un-ionized atoms are in the first and second states so $n_{Tot} = n_1 + n_2$, then it follows that

$$\frac{n_k}{n_{Tot}} = \frac{n_k}{n_1 + n_2} = \frac{\dfrac{n_k}{n_1}}{1 + \dfrac{n_2}{n_1}} = \frac{B(1,k)}{1 + B(1,2)} \qquad (12.5)$$

Equation 12.5 is in a useful form for us, since we can now express energy-level populations in terms of the Boltzmann function. We could simply set the total count to unity and then we would have population sizes as a fraction of the total. Or, we could simply use the total count as a scaling factor to control chart scales.

We are now ready to estimate our Balmer and Lyman spectral line intensities by:

1. Specifying a temperature T and electron pressure $P_e$.

2. Calculating the fraction of atoms that are ionized using the Saha equation (either Equation 12.2a or 12.2b).

3. Calculating the fraction (f) of atoms that are un-ionized from Equation 12.3.

4. Specifying a scaling using the total number of atoms $n_{Tot}$.

5. For a transition from level k down to level j, estimate the relative intensity as:

   $I_{kj} = n_k * f$ using Equation 12.5.

6. For Balmer transitions, use a lower level of j = 2, for Lyman use j=1.

Our models will rely on two main classes, **saha** and **balmer**, and a utility found online that was wrapped in a class wrapper for converting wavelength values into (R,G,B,A) colors. For completeness, the wavelength to color class is shown in Appendix II.

## CLASS SAHA PROGRAMMING NOTES

To support our models, a small Python class (**saha**) was developed to handle ionization-related calculations.

It relies on standard constants from Physics, but needs the partition numbers ZI and ZII, the electron pressure (**self.Pe**), and the ionization energy χ be specified. Calling (instantiating) the **saha** class creates lists of the ionization ratios (**self.NN**) and ionized (**self.NII_f**) fractions that can be used by its **self.plot_NII_fractions** function to show how the ionization fraction changes with temperature (see Figure 12.2) or be used to find the ionization ratio at a specific temperature with its **self.get_NN(t)** function. Note the ionization energy is in eV (electron volt) units, and the electron pressure normally ranges between 0.1 and 100.

The **saha** class is shown in Figure 12.3 and is very straightforward. If the class is run by itself, the stub at lines 61–64 will produce a plot like that in Figure 12.2.

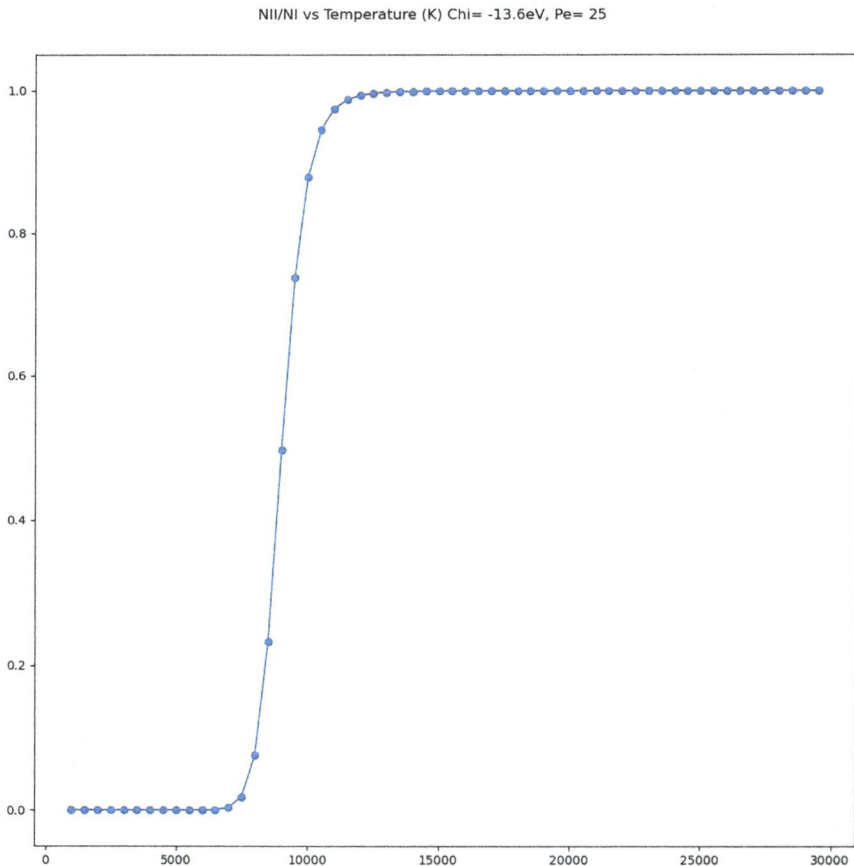

NII/NI vs Temperature (K) Chi= -13.6eV, Pe= 25

FIGURE 12.2 The ionization fraction for hydrogen as a function of temperature. Note how the fraction changes rapidly near a temperature of about 9600 K. Below 7000 K, the gas is mostly un-ionized, and above 12000 K, it is mostly ionized.

```
1. import math
2. import matplotlib.pyplot as plt
3.
4. class saha:
5.
6.     def __init__(self):
7.         self.NN      = []
8.         self.NII_f   = []
9.         self.Pe      = 25                    # [0.1, 100]
10.        self.ZI      = 0
11.        self.ZII     = 0
12.        self.c_saha  = 0
13.        self.T       = 0
14.        self.k       = 1.38e-23
15.        self.me      = 9.109e-31
16.        self.h       = 6.626e-34
17.        self.eV      = 1.602e-19
18.        self.chi     = -13.6                 # eV
19.        self.temps   = list(range(1000,30000, 500))
20.        self.get_partition_Zs()
21.        self.set_saha_constant()
22.        self.get_NN_ratios()
23.        self.get_NII_fractions()
24.
25.    def get_partition_Zs(self):
26.        self.ZI  = 2
27.        self.ZII = 1
28.
29.    def set_saha_constant(self):
30.        k  = self.k
31.        me = self.me
32.        pi = math.pi
33.        h  = self.h
34.        c1          = 2*k*self.ZII/self.ZI
35.        c2          = 2*pi*me*k/h/h
36.        self.c_saha = c1*(c2**1.5)
37.
38.    def get_NN(self,T):                       # Saha Fn - ionization ratio at T
39.        val = self.c_saha/self.Pe
40.        val = val*(T**2.5)
41.        e   = math.e
42.        val = val*pow(e, self.chi*self.eV/self.k/T)
43.        return val
44.
45.    def get_NN_ratios(self):                  # get ionization ratios for all T
46.        for T in self.temps:
47.            self.NN.append(self.get_NN(T))
48.
49.    def get_NII_fractions(self):              # get ionization fractions
50.        for r in self.NN:
51.            self.NII_f.append(r/(1+r))
52.
53.    def plot_NII_fractions(self):
54.        mydpi=100
55.        fig = plt.figure(figsize=(1200/mydpi,1200/mydpi),dpi=mydpi)
56.        txt = "NII/NI vs Temperature (K)"+" Chi= "+str(self.chi)+"eV"+", Pe= "+str(self.Pe)
57.        fig.suptitle(txt)
58.        plt.plot(self.temps,self.NII_f, '-o')
59.        fig.savefig('./Fig saha.jpg', dpi=mydpi)
60.
61. if __name__ == '__main__':
62.     sa = saha()
63.     sa.plot_NII_fractions()
64.
65.
```

FIGURE 12.3    Class saha.

To adjust the model change the **ZI** and **ZII** parameters with the **get_partition_Zs** function and set the electron pressure and the partition numbers (lines 16 and 25). To support plotting with the **plot_NII_fractions()** function, lists of the ionization ratios **self.NN** and of the un-ionized fractions for different temperatures are created by the **get_NN_ratios** and **get_NII_fractions** functions. The ionization ratio for a given temperature can be calculated using the **get_NN()** function.

## SIMULATING THE BALMER SPECTRUM

The **balmer** class is the one which will model the Balmer lines, plot the line intensities, and a synthetic spectrum. A short program (**test_balmer.py**) to demonstrate how it works is shown in Figure 12.4.

The **test_balmer.py** program will create plots for line intensities and synthetic spectra similar to those shown in Figures 12.5 and 12.6.

In Figure 12.5, the intensities are dominated by the red hydrogen-alpha line at low temperatures, and the strongest effect is near 9000 K, but the overall intensity decreases with higher temperatures as the majority of atoms become ionized.

In Figure 12.6, synthetic spectra generated by our models with the lines normalized to the peak intensity show how as the temperature increases,

```
 1. import matplotlib.pyplot as plt
 2. from balmer import balmer
 3.
 4. sp = balmer()
 5.
 6. plt.rcParams['axes.facecolor'] = 'white'
 7. sp.plot_NII_fractions()
 8. temps      = [5000,7000,9000,20000]
 9. nsubplots = len(temps)
10.
11. # plot line amplitudes
12.
13. mydpi=100
14. fig = plt.figure(figsize=(1200/mydpi,1200/mydpi),dpi=mydpi)
15. plt.rcParams['axes.facecolor'] = 'white'
16. txt = "Balmer Line Intensities vs Wavelength (nm), Pe="+str(sp.Pe)+", x"+str(sp.scale)
17. fig.suptitle(txt)
18. for i in range(0,nsubplots):
19.     plt.subplot(nsubplots, 1, i+1)
20.     sp.get_balmer_lines(temps[i])
21.     sp.plot_balmer_lines()
22. plt.savefig('./Fig Balmer Intensities.jpg',dpi=mydpi)
23.
24. # plot synthetic spectrum
25.
26. fig = plt.figure(figsize=(1200/mydpi,1200/mydpi),dpi=mydpi)
27. plt.rcParams['axes.facecolor'] = 'black'
28. txt = "Balmer Synth. Spectrum, Pe="+str(sp.Pe)
29. fig.suptitle(txt)
30. for i in range(0,nsubplots):
31.     plt.subplot(nsubplots, 1, i+1)
32.     sp.get_balmer_lines(temps[i])
33.     sp.plot_balmer_spectrum()
34. plt.savefig('./Fig Balmer synth.jpg',dpi=mydpi)
35. plt.show()
36.
```

FIGURE 12.4   A short program (**test_balmer.py**) to plot the ionization fraction, and the Balmer line intensities and a synthetic spectrum for a set of listed temperatures (see line 8).

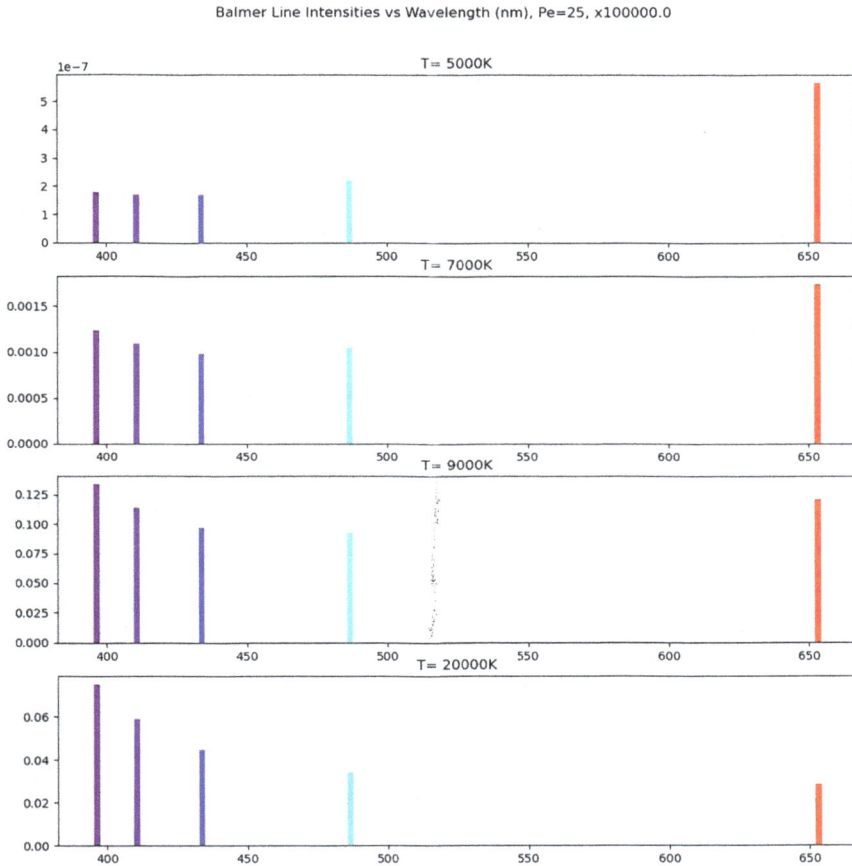

FIGURE 12.5 Our model's Balmer line intensities for four different temperatures.

more of the higher-level transitions become visible near the blue/ultraviolet short wavelength. The overall series intensities are weaker at high temperatures.

## CLASS BALMER PROGRAMMING NOTES

The complete code listing of class **balmer** is shown in Figure 12.7.

The **balmer** class uses both the **saha** and **w2rgb** classes and is intended to produce charts showing the line intensities and synthetic spectra for a selection of temperatures and a specified electron pressure. The synthetic spectra use a fixed line width, and to emulate intensity variation, without changing the color, the RGB triplet component values for a line's wavelength are each multiplied by the ratio of the initial line intensity to the

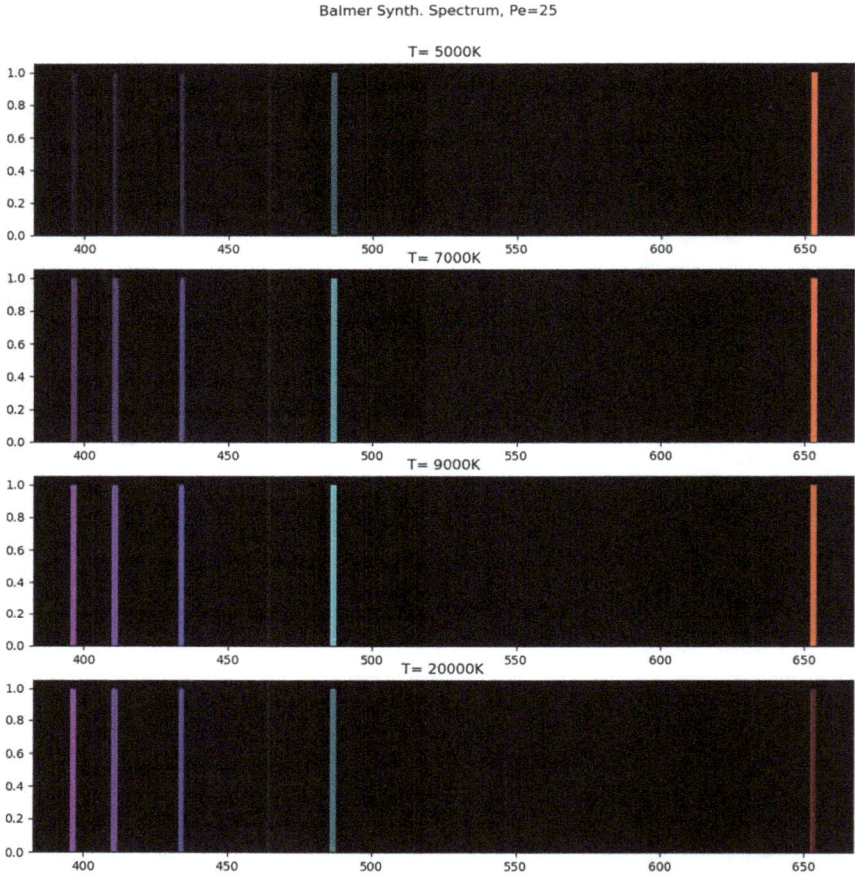

FIGURE 12.6   The corresponding synthesized spectra for the results in Figure 12.5.

series maximum for that temperature. As described earlier, line intensity is estimated from the population at the upper level $n_1$, multiplied by the fraction of neutral atoms present.

Note, in the code, we use a naming convention where emission lines occur between a higher level 'k,' and a lower one 'j,' where for the Balmer series, $j = 2$. If other spectral series are of interest, **j** should be changed, for example, for Lyman, use **j** = 1.

Class **balmer** consists of three main function groups. The first group is for initializing the class, and also provides some utility function for calculating the wavelength (**get_lambda**) associated with an electron falling from level $n_1$ to level $n_0$ in the hydrogen atom.

```
1. import math
2. import matplotlib.pyplot as plt
3. from saha import saha
4. from wavelength2rgb import w2rgb
5.
6. class balmer(saha,w2rgb):
7.
8.     def __init__(self):
9.         super().__init__()
10.        self.g          = []
11.        self.scale      = 1e5                    # arbitrary scaling for charts
12.        self.Pe         = 25                     # [0.1, 100]
13.        self.T          = 0
14.        self.nn         = []
15.        self.k          = 1.38e-23
16.        self.eV         = 1.602e-19
17.        self.c          = 3e8
18.        self.E          = [-13.6, -3.4, -1.5, -0.85, -0.54, -0.38, -.27]
19.        self.nlevels    = len(self.E)
20.        self.n          = list(range(1,self.nlevels+1,1))
21.        self.linewidth = 2
22.        self.balmer     = []
23.        self.lambdas    = []
24.        self.bcolor     = []
25.        self.bw         = []
26.        self.lambdas    = []
27.        self.set_g_values()
28.
29.    def get_E(self,l):                           # Want get_E(1) to return E[0] etc.
30.        return self.E[l-1]
31.
32.    def get_g(self,indx):
33.        return self.g[indx-1]
34.
35.    def get_lambda(self, j,k):
36.        E_diff = (self.get_E(k) - self.get_E(j)) *self.eV
37.        f      = E_diff/self.h
38.        wlen   = self.c/f
39.        wlen   = wlen*1e9
40.        return wlen
41.
42.    def set_g_values(self):                      # calculate hydrogen degeneracies
43.        for i in  self.n:
44.            self.g.append(2*i*i)
45.
46.    def get_nn(self,j,k,t):                  # get level ratios from Boltzmann eqn
47.        e = math.e
48.        gj = self.get_g(j)
49.        gk = self.get_g(k)
50.        del_E = (self.get_E(k) - self.get_E(j))*self.eV   #n0=2 - doing Balmer
51.        val = pow(e,-del_E/self.k/t)*gk/gj
52.        return val
53.
54.    def get_n_k(self,k,t):                       # atoms at level k (assumes n_tot=n1+n2)
55.        bk = self.get_nn(1,k,t)
56.        b2 = self.get_nn(1,2,t)
57.        nk = bk/(1+b2)
58.        nk = nk*self.scale
59.        return nk
60.
61.    def get_single_line_I(self,t,j,k):
62.        I = []
63.        r = self.get_NN(t)
64.        f = r/(1+r)
65.        print(t,j,k)
66.        nk=self.get_n_k(k,t)
67.        I = (nk*(1-f))/(j-1)          # assuming 1/(n1-1) electrons fall into n0
68.        wlen = self.get_lambda(j,k)
69.        color = self.w2rgb(wlen)
70.        return  wlen, I, color
71.
72.    def get_balmer_lines(self,t):
73.        self.balmer  = []
74.        self.lambdas = []
75.        self.bcolor  = []
```

FIGURE 12.7    Class **balmer**.                                    (*Continued*)

```
76.          self.bcolorI = []   #color and intensity
77.          self.bw     = []
78.          self.T      = t
79.
80.          Imax = 0
81.          for i in range(3,self.nlevels+1):
82.              wlen,I,bcol = self.get_single_line_I(t,2,i)
83.              if I > Imax:
84.                  Imax = I
85.
86.              self.balmer.append(I)
87.              self.lambdas.append(wlen)
88.              self.bw.append(self.linewidth)
89.              self.bcolor.append(bcol)
90.
91.          for i in range(0,self.nlevels-2):       # adjust color brightness
92.              iratio    = self.balmer[i]/Imax
93.              (r,g,b,a) = self.bcolor[i]
94.              r2        = r*iratio
95.              g2        = g*iratio
96.              b2        = b*iratio
97.              col2      = (r2,g2,b2,a)
98.              self.bcolorI.append(col2)
99.
100.     def plot_balmer_lines(self):
101.         txt2 = 'T= '+str(self.T)+'K'
102.         plt.title(txt2)
103.         plt.bar(self.lambdas, self.balmer, color = self.bcolor,width=2)
104.
105.     def plot_balmer_spectrum(self):
106.         txt2 = 'T= '+str(self.T)+'K'
107.         plt.title(txt2)
108.         y = len(self.lambdas)*[1]
109.         print(self.bw)
110.         plt.bar(self.lambdas, y, color = self.bcolorI,width=self.bw)
111.
```

FIGURE 12.7 (CONTINUED)   Class **balmer**.

Because Python indexing starts from zero, but the usual labelling in physics for hydrogen levels starts with one, the **get_E()** and **get_g()** functions allow us to use the physics indexing with the energy **E** and degeneracy **g** lists.

The second group of functions use the Boltzmann and related equations to calculate the population density ratio (**self.get_nn**), the number of atoms in state k (**self.get_n_k**), the intensity of a transition from **k** down to **j** (**self.get_single_line_I**), and a function to generate the line intensities for the Balmer sequence (**self.get_balmer_lines**).

And finally, there are two plotting functions to create line intensity charts and synthetic spectra.

## CHAPTER 13

# Isochrons

## *The Ages of Rocks*

I N BASIC PHYSICS COURSES, we learn about radioactive decay, where some 'parent' atoms change into different ones – 'daughters.' This can happen when protons change into neutrons, or when alpha particles with two protons and two neutrons are emitted, and since each element has a certain number of protons in its nucleus (its atomic number), changing the number of protons changes it into another element. If we knew how many parent atoms were present initially in a sample when it formed, then the number of daughter atoms would allow us to calculate the age of the sample, since decay rates are well documented. However, if the initial sample had a mix of daughter and parent elements when it formed, the age estimate is more difficult to determine. In this chapter, we will explore the isochron method, which is a way to address the problem of daughter atoms being present when rock forms and show how the technique can yield accurate age estimates.

We will often refer to the sample being investigated as being a crystal or mineral sample, since, because of their nature, crystals have their compositions set initially and can only change if trapped/embedded atoms undergo radioactive decay.

Radioactive decay is a random process where atoms of the 'parent' element transform into the 'daughter' element. We cannot know when a

DOI: 10.1201/9781003600046-13

particular atom will transform, but we can make accurate estimates of how fast they are changing, and through study, determine the half-life 'tau,' for example, the time needed for half of the current number of parent atoms to transform. If we started with 64 parent atoms, after one half-life, 32 would remain; after another half-life 16 of the original 64 parent atoms would be left and so on. Half-lives can range from $10^{-24}$ to $10^{+22}$ seconds, depending on the element.

(It is important to note that changing the number of protons can make a huge difference in the atom's chemical properties – essentially how it behaves and interacts with other elements – because changing the proton count will change the number of attached electrons. For example, changing a proton in nitrogen would result in carbon – an element with very different physical properties. On the other hand, isotopes of an element have the same number of protons and electrons, but due to different numbers of neutrons, their chemical properties remain unchanged.)

Mathematically we can say that if there are $P_0$ parent atoms initially of an element with a half-life $\tau$, then after time t, the remaining number of atoms is:

$$P = P_0 e^{\frac{-0.693t}{\tau}} = P_0 e^{-\lambda t} \tag{13.1a}$$

where $\lambda$ (the 'decay constant') is **ln(2)/τ** which is approximately **0.693/τ**.

After one half-life (t = τ) P = $P_0$/2 since $e^{-0.693}$ = ½.

Note we could express the number of radiogenic daughters produced by the decay as:

$$P0 - P(t) = P_0 \left(1 - e^{-\lambda t}\right) \tag{13.1b}$$

and it follows that

$$\frac{\Delta P}{P(t)} = \left(e^{\lambda t} - 1\right) \tag{13.1c}$$

Alternatively, we could write:

$$P = P_0 \left(\frac{1}{2}\right)^{\frac{t}{\tau}} \tag{13.2}$$

Obviously, the number of daughter atoms present must be $D = P_0 - P$ if the initial sample only had parent atoms, otherwise,

$$D = P_0 - P + D_0 \qquad (13.3)$$

where $D_0$ is the initial amount present at the mineral formation. If $D_0$ is zero (i.e., there was no daughter material present when the rock formed,) then by measuring $D$ and $P$, we could figure out the sample age using either Equation 13.1 or 13.2. These equations allow $t$ to be expressed in terms of 3 variables ($P$, $P_0$, and $\tau$), so measuring the number of $D$ and $P$ atoms, from which $P_0$ could be estimated, means all three variables would be known, assuming $\tau$ is already known from lab experiments. However, assuming the sample had no daughter material present initially is not necessarily true, which means our age determination task has to deal with four variables, not three, we need something else, and this is where the isochron technique can be used.

In this chapter, we will explain and model the isochron technique which is designed to deal with daughter atoms in the initial sample and show how it can be used to estimate ages of rocks.

## ESTIMATING THE AGE OF ROCKS USING ISOCHRONS

The isochron method relies on analyzing amounts of different atom types and it is important we have a clear notation to minimize confusion. There are three kinds of atoms we will track in our analysis. First there is the parent type, $P$, and this radioactively decays into a daughter type. For the isochron techniques, there must be two daughter isotopes; $R$ is the radiogenic one, where $P$ decays into $R$, and $I$ is an inert isotope of $D$. $R$ and $I$ are indistinguishable chemically, and in any mineral, there will be a certain number of daughters for every parent, and this ratio will probably be different for a different mineral.

We will assume there are no processes that can change $I$, and so the amount of $I$ in a rock or a sample remains unchanged after formation. When thinking in terms of samples from a rock, we are thinking of crystals of different minerals, that once set, lock in the atoms, so there is no change from atoms entering or leaving the crystal.

Let's now suppose there is a radioactive element ($P$, the parent) that decays to become a daughter element $R$. A mineral might have a particular ratio of $P$ to $D$ atoms; it doesn't care about the proportions of $R$ and $I$

– only that their sum is in the correct proportion to **P**. (To keep things simple, we will use 'P' to refer to the atom type or to the number of those atoms, and similarly for **R** and **I**.)

Minerals/crystals are the result of chemical processes, and, as we said, do not care about the isotope differences. For example, each 100 atoms of a mineral might need 20 **P** and 10 **D** types of atoms, and it doesn't matter if the daughter atoms consist of 8 or 3, or whatever, of the radiogenic, as long as **R** and I add up to 10.

A different mineral, for every 100 atoms, might need 40 **D** and 8 **P** atom types, and again, it doesn't matter how much of **R** and **I** is in the 40 **D** atoms, just as long as they add up to 40, for that mineral.

Now, let's suppose a rock is forming from a molten mass, and during its formation, crystals of both minerals are forming, and will become crystals in the rock once it has formed. Since both types of crystals formed from the same initial molten mass, we can reasonably assume both minerals will have the same ratios for their initial **R** and **I** isotopes.

For example, let's assume **R** and **I** are equally likely in the molten mass and a rock forms with crystals of both mineral types. Then, for every 100-atom sample from the first mineral, there might be 20 **P**, 5 **R**, and 5 **I**, and for the second mineral 8 **P**, 20 **R**, and 20 **I** because at **t = 0**, when the rock formed, **R = I**. However, as time passes, and parent atoms turn into **R** atoms, after one half-life, the crystal sample compositions will have (10 **P**, 15 **R**, 5 **I**) and (4 **P**, 24 **R**, 20 **I**) respectively, so the **R** to **I** ratio changes with age, and is different, for each mineral. This difference in behavior represents a new aspect that can be measured, that is used as a 'known' by the isochron technique, to correct for the 'unknown' initial daughter quantity.

If we could monitor a crystal sample over time, **P** and **R** would change, but **I** would remain constant, and all would depend on the sample size. To exclude the effects of sample size, let's normalize all the **P** and **R** measurement, by scaling them to **I**, in other words, we will consider the quantities **P/I** and **R/I** (as determined for each different element) as being of primary interest. Note also that when determining the amounts of atoms present using mass spectroscopy, it is easier to compare two types than to get estimates of their individual quantities, which is another very important reason to work with these ratios.

For our rock, we can plot our measurement on a chart where the X-axis is **P/I** and the Y-axis is **R/I**. There will be a (**P/I**, **R/I**) data point from each mineral in the rock. If, for our example, we plotted two points, one from each mineral, we would find the line through them, called an isochron, would intercept the Y-axis at a particular **R/I** value. As we shall demonstrate, all the isochrons, for all **t** values studied, will all intersect the Y-axis at the same value, and this must be the initial (**t = 0**) **R/I** value since that is the only **R/I** value they all have in common. Knowing this, we can now say what the initial **R** was, and figure out the elapsed time, and hence the rock's age.

We can express the radiogenic daughter count at time **t** as the sum of the initial count and the count of those subsequently created from parent atoms are decaying.

$$R(t) = R(0) + \Delta P(t) \tag{13.4}$$

where $\Delta$**P(t)** is the number of daughters created by the decay.

Rewriting as:

$$R(t) = R(0) + \frac{\Delta P(t)}{P(t)} P(t) \tag{13.5}$$

and then rescaling by **I**, and using Equation 13.1c yields:

$$\frac{R(t)}{I} = \frac{R(0)}{I} + \frac{\Delta P(t)}{P(t)} \frac{P(t)}{I} \tag{13.6}$$

$$= \frac{R(0)}{I} + \left(e^{\lambda t} - 1\right) \frac{P(t)}{I} \tag{13.7}$$

In this form, a plot of measured (current) quantities **R(t)/I** against **P(t)/I** has slope

$$m = \left(e^{\lambda t} - 1\right) \tag{13.8}$$

and an intercept **R(0)/I**. So now our problem is solved: We now know the initial count of daughter atoms and can use the slope to estimate the age:

$$t = \frac{\ln(m+1)}{\lambda} \tag{13.9}$$

A short program (**test_isochron.py**) demonstrating the isochron class is shown in Figure 13.1. After running **test_isochron.py**, the chart shown

```
 1. import isochron
 2. from tabulate import tabulate
 3.
 4. iso            = isochron.isochron()
 5. iso.tau        = 1
 6. iso.I_fraction = 0.5
 7. iso.times      = [0,1,2,4,8]
 8. P              = 1000
 9. PD_ratio       = 2
10. RD_ratio       =.5    # must be same for all minerals from same rock
11. iso.x0         = iso.make_crystal('sample1', P, PD_ratio, RD_ratio)
12. P              = 2000
13. PD_ratio       = 3
14. iso.x1         = iso.make_crystal('sample2', P, PD_ratio, RD_ratio)
15.
16. iso.plot_isochrons()
17.
18.
19. head= ['t','Sample',"P","D","R","I","R/I","P/I"]
20. print(tabulate(iso.data, headers=head, tablefmt="grid", numalign='right'))
21.
```

FIGURE 13.1   A short code (**test_isochron.py**) that generates two mineral types and draws isochrons for different elapsed times. It generates a plot and also a table which is sent to the console.

in Figure 13.2 is produced, and printed output is sent to the console (see Figure 13.3).

Two sample minerals were defined using the **make_crystal()** routine (see lines 11 and 14). The first sample had 1000 parent (**P**) atoms with 500 daughter (**D**) atoms since the parent-daughter ratio (**PD_ratio**) was set to 2 (line 9). The ratio of **R** to **D** daughter isotopes was set to **RD_ratio** = .5 at line 10, so there were 250 **R** and **I** = **D**−**R** (250) inert isotopes. In this model, the **P:D** ratio is dictated by the mineral's chemistry at the time of formation, the while different minerals can have different **PD_ratio**s, all must have the same **RD_ratio**.

At any time t, **get_crystal_counts_at_time_t()** returns the population counts of the parent and daughter atoms in the specified crystal.

The isochron chart shows that the slope is zero at **t** = **0**, as expected by setting **t** = **0** in Equation 13.8. The tabular output is useful for seeing how the counts of various atoms change among samples and for different times. For example, the **t** = **0** isochron (blue) connects the points (**R/I**, **P/I**) for samples: (250/250, 1000/250) for sample 1 on the left end and (333/333, 2000/333) for sample 2 on the right end, or equivalently (4,1) to (6,1). The y-intercept is 1 which is correct since **R/I** is 1 at t = 0.

Overall, the model has worked well and has demonstrated important aspects such as all isochrons having the same intercept, and how the age can be estimated from their slopes. Figure 13.3 also shows the estimated ages displayed to the console. In this case, they are extremely accurate

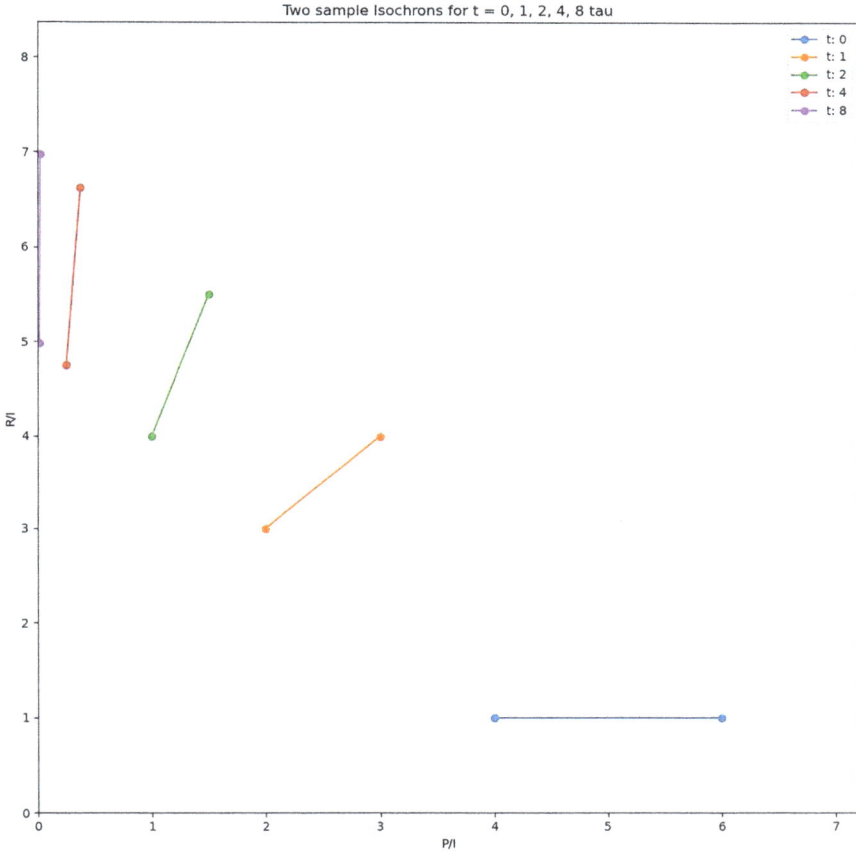

FIGURE 13.2  Isochrons for the two samples, evaluated at three different elapsed times. The slope of an isochron is used to estimate the isochron's age. Note how the isochrons all point back to the same intercept.

since there is no noise in our models. In reality, there would be uncertainties in the various measurements and so the isochrons would be less precise and hence the age estimates.

## CLASS ISOCHRON PROGRAMMING NOTES

Class isochron (see Figure 13.4) was written to explore the isochron method where different half-lives and element compositions can be specified. Mineral compositions are set by defining a parent-to-daughter ratio (**PD_ratio**), and, very importantly, the fractions of radiogenic and inert daughter isotopes using the **RI_ratio** parameter.

```
 1. +++++++++++++++++++++++++++++++++++++++++++++
 2. slope, intercept =  0.0 1.0
 3. R0_est            =  250.0 R1_est =  333.3333333333333
 4. Estimated age     =  0.0  tau
 5. +++++++++++++++++++++++++++++++++++++++++++++
 6. slope, intercept =  1.0 1.0
 7. R0_est            =  250.0 R1_est =  333.3333333333333
 8. Estimated age     =  1.0  tau
 9. +++++++++++++++++++++++++++++++++++++++++++++
10. slope, intercept =  3.0 1.0
11. R0_est            =  250.0 R1_est =  333.3333333333333
12. Estimated age     =  2.0  tau
13. +++++++++++++++++++++++++++++++++++++++++++++
14. slope, intercept =  15.0 1.0
15. R0_est            =  250.0 R1_est =  333.3333333333333
16. Estimated age     =  4.0  tau
17. +++++++++++++++++++++++++++++++++++++++++++++
18. slope, intercept =  255.0 1.0
19. R0_est            =  250.0 R1_est =  333.3333333333333
20. Estimated age     =  8.0  tau
21. +-----+---------+---------+---------+---------+---------+
22. |  t  | Sample  |    P    |    D    |    R    |    I    |
23. +=====+=========+=========+=========+=========+=========+
24. |  0  | sample1 |   1000  |    500  |    250  |    250  |
25. +-----+---------+---------+---------+---------+---------+
26. |  0  | sample2 |   2000  | 666.667 | 333.333 | 333.333 |
27. +-----+---------+---------+---------+---------+---------+
28. |  1  | sample1 |    500  |   1000  |    750  |    250  |
29. +-----+---------+---------+---------+---------+---------+
30. |  1  | sample2 |   1000  | 1666.67 | 1333.33 | 333.333 |
31. +-----+---------+---------+---------+---------+---------+
32. |  2  | sample1 |    250  |   1250  |   1000  |    250  |
33. +-----+---------+---------+---------+---------+---------+
34. |  2  | sample2 |    500  | 2166.67 | 1833.33 | 333.333 |
35. +-----+---------+---------+---------+---------+---------+
36. |  4  | sample1 |   62.5  |  1437.5 |  1187.5 |    250  |
37. +-----+---------+---------+---------+---------+---------+
38. |  4  | sample2 |    125  | 2541.67 | 2208.33 | 333.333 |
39. +-----+---------+---------+---------+---------+---------+
40. |  8  | sample1 | 3.90625 | 1496.09 | 1246.09 |    250  |
41. +-----+---------+---------+---------+---------+---------+
42. |  8  | sample2 |  7.8125 | 2658.85 | 2325.52 | 333.333 |
43. +-----+---------+---------+---------+---------+---------+
44.
```

FIGURE 13.3    Console output showing the slopes, intercepts, and estimate ages for the different isochrons, and the **P**, **D**, **R**, and **I** atom counts for the two samples at different ages. This output is useful for displaying counts at various stages of the analysis that could be cross checked with separate calculations.

To create the isochron chart, the physical parameters are specified (populations, half-life, mineral composition) and the **plot_isochrons()** function invoked to plot the isochron for each requested time in the **self. times[]** list (see the example at lines 106–121).

An isochron is generated by the **get_isochron** function, which uses the counts from each crystal at time t to calculate the ratios needed for isochron line drawing (see lines 49 and55). Obviously, many different models could be tested here by changing the **PD_ratio** and **RI_ratio** ratios and the number of initial parent atoms in a crystal/sample; the small number of atoms was chosen for simplicity and to avoid having to use scientific notation in data results.

```
1.  import math
2.  import matplotlib.pyplot as plt
3.  from tabulate import tabulate
4.
5.  class isochron:
6.
7.      def __init__(self):
8.          self.data       = []
9.          self.tau        = 1
10.         self.times      = [0,0.5, 1.5, 2.5]
11.
12.     def make_crystal(self,label,P,PD_ratio, RD_ratio):
13.         D  = P/PD_ratio
14.         R  = D * RD_ratio
15.         I  = D - R
16.         xIct = {
17.         'type'      : label,
18.         'P'         : P,            # Number of parent atoms P
19.         'PD_ratio' : PD_ratio,       # Parents per daughter in mineral
20.         'D'         : D,             # Number of daughters in sample
21.         'I'         : I,             # Num. of non-radiogenic daughters
22.         'R'         : R              # Number of radiogenic daughters
23.         }
24.         return xIct
25.
26.     def get_crystal_counts_at_time_t(self,crystal,t):
27.         P       = crystal['P']*(0.5 **(t/self.tau))
28.         del_p   = crystal['P'] - P      # number of decays
29.         D       = crystal['D'] + del_p
30.         R       = crystal['R'] + del_p
31.         I       = crystal['I']
32.          return [t,crystal['type'],P,D,R,I]
33.
34.     def get_initial_crystal_values(self):
35.         counts = self.get_crystal_counts_at_time_t(self.x0,0)
36.         [t1,name1,p1,d1,R1,I1] = counts
37.         print('x0 Initial values: ',p1,d1,R1,I1)
38.         self.data.append(counts)
39.
40.         counts = self.get_crystal_counts_at_time_t(self.x1,0)
41.         [t2,name2,p2,d2,R2,I2] = counts
42.         print('x1 Initial values: ',p2,d2,R2,I2)
43.         self.data.append(counts)
44.
45.     def get_isochron(self,t):
46.
47.         print('+++++++++++++++++++++++++++++++++++++++++++')
48.
49.         counts= self.get_crystal_counts_at_time_t(self.x0,t)
50.         self.data.append(counts)
51.         [t,type0,P0,D0,R0,I0] = counts
52.         x0 = (P0/I0)
53.         y0 = (R0/I0)
54.
55.         counts = self.get_crystal_counts_at_time_t(self.x1,t)
56.         self.data.append(counts)
57.         [t,type1,p1,d1,R1,I1] = counts
58.         x1 = (p1/I1)
59.         y1 = (R1/I1)
60.
61.         slope = round((y1-y0)/(x1-x0),4)
62.         icpt  = round(y1-slope*x1, 4)            # R/I at t=0
63.         R0_est = icpt*I0         # Est. init. rad. daughters in sample 0
64.         R1_est = icpt*I1         # Est. init. rad. daughters in sample 1
65.         lambda0 = math.log(2)/self.tau
66.         age = round(math.log(slope+1)/lambda0,2)
67.         print('slope, intercept = ',slope,icpt)
68.         print('R0_est     = ', R0_est, 'R1_est = ', R1_est)
69.         print("Estimated age     = ",age, " tau")
70.         return [x0,x1], [y0,y1]
71.
72.     def plot_isochrons(self):
73.         mydpi=100
74.         fig = plt.figure(figsize=(1200/mydpi,1200/mydpi),dpi=mydpi)
```

FIGURE 13.4 (CONTINUED)   Class isochron.

*(Continued)*

```
75.         ax = plt.gca()
76.
77.         tstr = ""
78.         xs   = []; ys   = []        # save x and y vals to set plot max/min
79.         for t in self.times:
80.             tstr = tstr +str(t)+', '
81.             lstr = "t: "+str(t)
82.             x,y  = self.get_isochron(t)
83.             xs   = xs+x
84.             ys   = ys+y
85.             plt.plot(x,y,'-o',label=lstr)
86.
87.         xmax = 1.2*max(xs)
88.         ymax = 1.2*max(ys)
89.
90.         plt.xlim(0,xmax)
91.         plt.ylim(0,ymax)
92.         legend=plt.legend()
93.         frame=legend.get_frame()
94.         frame.set_facecolor('w')
95.         plt.xlabel("P/I")
96.         plt.ylabel("R/I")
97.         tstr = tstr[:-2]
98.         plt.title("Two sample Isochrons for t = "+tstr+ " tau")
99.         ax.set_facecolor('w')
100.        #plt.show()
101.        plt.savefig('./Fig isochrons.jpg',dpi=mydpi)
102.
103. if __name__ == '__main__':
104.
105.        iso            = isochron()
106.        iso.tau        = 1
107.        iso.I_fraction = 0.5
108.        iso.times      = [0,1,2,4,8]
109.        P              = 1000
110.        PD_ratio       = 2
111.        RD_ratio       =.5   # must be same for all minerals from same rock
112.        iso.x0         = iso.make_crystal('sample1', P, PD_ratio, RD_ratio)
113.        P              = 2000
114.        PD_ratio       = 3
115.        iso.x1         = iso.make_crystal('sample2', P, PD_ratio, RD_ratio)
116.
117.        iso.plot_isochrons()
118.
119.
120.        head= ['t','Sample',"P","D","R","I","R/I","P/I"]
121.        print(tabulate(iso.data, headers=head, tablefmt="grid", numalign='right'))
122.
```

FIGURE 13.4 (CONTINUED)   Class isochron.

## SUMMARY

The isochron method is a very powerful technique that relies on the presence of a stable daughter isotope to provide an additional feature that can be measured. It assumes all minerals, even with daughter atoms present, must have a fixed daughter isotope ratio when the mineral was formed. This ratio changes as minerals age, and parent atoms decay into daughters, so even though different minerals will have different parent to daughter ratios, even at t = 0, all minerals share the property that their daughter isotope ratio is the same for all initially.

In this chapter, we provided a simple code to model the technique and found the estimated ages derived from the isochron method did indeed match the model ages.

Model improvements that could be explored would include using experimentally determined isotope ratios and atom counts, and perhaps adding a noise component to the simulation to study the technique's robustness; for example, beyond what age do slope estimates become unacceptably imprecise? It would also be easy (desirable) to add additional crystals/minerals to the sample set, to produce isochrons with more than two points!

# Appendices

## APPENDIX I: CLASS SPATIAL

### Class Spatial Programming Notes

Class spatial was developed to emphasize the spatial relationships between stars by specifying a radius, identifying all neighbors within that radius for all stars, and drawing radials from each star to a fixed number of neighbors.

An example scenario is shown following the usual '**if __name__ ==**' construct at line 81.

The first 40 stars (the brightest) are read in from the catalog using **read_catalog()** on line 89, and **get_ijxy()** is used to extract matrix and cartesian versions of the star coordinates (line 91), after which a KDTree (sp.T) is built (line 92). The **KDTree** structure creates a balanced tree that facilitates searching for neighbor.

Function **make_plots()** creates the two panel output chart and specifies the maximum number of radials to be drawn for any star (lines 97 and 69).

Radials are plotted by **plot_all_radials()** which, for every star (line 48), searches for neighbors within distance r = 300 and plots that star's radials (up to **Nmax** of these) (lines 50–51). Each star's radials has the same color, and colors are selected from a list **self.pcolors** (line 13).

## APPENDIX II CONVERTING WAVELENGTH TO COLOR

The **w2rgb** class converts wavelengths expressed in nm (nanometers) to color (RGBA) – which supports creation of nice charts and the synthetic spectra. It uses a function written by Dan Bruton as described in the comments included with the function.

Since we use the code simply for graphical and visual effect, we accept it as being suitable for our needs because its results are visually effective and the code is clear and easy to follow:

```python
 1. import csv
 2. import numpy as np
 3. import matplotlib.pyplot as plt
 4. from scipy.spatial import KDTree
 5.
 6. class spatial:
 7.     def __init__(self):
 8.         self.proj_dir      = ''
 9.         self.fname         = ''
10.         self.star_cat1     = []
11.         self.ij            = []
12.         self.xy            = []      # used for screen plots
13.         self.pcolors       = ['b','c','g','k','m','r','y']
14.         self.path          = ''
15.         self.nrows         = 0
16.
17.     def read_catalog(self, N):
18.         line_count = 0
19.         star_cat_in = []
20.         with open(self.fpath, newline='') as csvfile:
21.             lines_in = csv.reader(csvfile, delimiter=',', quotechar='|')
22.             for row in lines_in:
23.                 line_count += 1
24.                 rowf = [float(i) for i in row]
25.                 star_cat_in.append(rowf)
26.                 if line_count > N:
27.                     break
28.         return star_cat_in
29.
30.     def get_ijxy(self,c,N):              # return matrix and cartesian coords
31.         for r in range(0,N):
32.             [i0,j0] = c[r][2:4]
33.             x0 = j0
34.             y0 = self.nrows - i0
35.             self.ij.append([i0,j0])
36.             self.xy.append([x0,y0])      # use x-y for screen plots
37.
38.     def find_nbrs(self,i,j,c,T,r):
39.         pts = np.array(c)
40.         idx = T.query_ball_point([i,j],r)
41.         return pts[idx]
42.
43.     def plot_all_radials(self,Nmax):
44.         r = 300                          # search radius
45.         p = 0
46.         count = 0
47.         c = self.xy
48.         for s in c:
49.             [x,y] = s
50.             nbr_list=self.find_nbrs(x, y, c, self.T, r)
51.             self.plot_radials(x,y, nbr_list ,Nmax,p)
52.             p += 1
53.             p = p % 7
54.             count += 1
55.
56.     def plot_radials(self,i,j,nlist,Nmax,p):
57.         count = 0
58.         for n in nlist:
59.             count += 1
60.             [i2,j2] = n
61.             di = i2 - i
62.             dj = j2 - j
63.             x = [i,i+di/3]
64.             y = [j,j+dj/3]
65.             plt.plot(x,y,linewidth=1,color = self.pcolors[p%7])
66.             if count > Nmax:                    #   plot first Nmax radials
67.                 break
68.
```

```
69.    def make_plots(self,x,y,Nmax):
70.        Nmax  = 6                    # max number of radials to draw
71.        mydpi=120
72.        fig = plt.figure(figsize=(1200/mydpi,1000/mydpi),dpi=mydpi)
73.        plt.subplot(1,2,1)
74.        plt.plot(x, y, 'o', color='black', markersize=2)
75.        plt.subplot(1,2,2)
76.        plt.plot(x, y, 'o', color='black', markersize=2);
77.        sp.plot_all_radials(Nmax)
78.        plt.show()
79.        plt.savefig('./Fig Spatial.jpg',dpi = mydpi)
80.
81. if __name__ == '__main__':
82.
83.    sp           = spatial()
84.    N            = 40                      # catalog entries to use
85.    sp.proj_dir = './'
86.    sp.fname    = 'T_CrB_20240512'
87.    sp.fpath    = sp.proj_dir  + sp.fname +'_cat.csv'
88.    sp.nrows    = 1720                     # Seestar 50 image
89.    sp.star_cat = sp.read_catalog(N)
90.
91.    sp.get_ijxy(sp.star_cat,N)
92.    sp.T        = KDTree(sp.xy)
93.
94.    x = [s[0] for s in sp.xy]              # star [x,y]
95.    y = [s[1] for s in sp.xy]
96.
97.    sp.make_plots(x,y,6)
98.
```

```
8.  ''' taken from http://www.noah.org/wiki/Wavelength_to_RGB_in_Python
9.  This converts a given wavelength of light to an
10. approximate RGB color value. The wavelength must be given
11. in nanometers in the range from 380 nm through 750 nm
12. (789 THz through 400 THz).
13.
14. Based on code by Dan Bruton
15. http://www.physics.sfasu.edu/astro/color/spectra.html
16. Additionally alpha value set to 0.5 outside range
17. '''
18.
19.
20. class w2rgb:
21.     def __init__(self):
22.         pass
23.
24.     def w2rgb(self,wavelength, gamma=0.8):   # Based on code by Dan Bruton.
25.
26.         wavelength = float(wavelength)
27.         if wavelength >= 380 and wavelength <= 750:
28.             A = 1.
29.         else:
30.             A=0.5
31.         if wavelength < 380:
32.             wavelength = 380.
33.         if wavelength >750:
34.             wavelength = 750.
35.         if wavelength >= 380 and wavelength <= 440:
36.             attenuation = 0.3 + 0.7 * (wavelength - 380) / (440 - 380)
37.             R = ((-(wavelength - 440) / (440 - 380)) * attenuation) ** gamma
38.             G = 0.0
39.             B = (1.0 * attenuation) ** gamma
40.         elif wavelength >= 440 and wavelength <= 490:
```

```
41.              R = 0.0
42.              G = ((wavelength - 440) / (490 - 440)) ** gamma
43.              B = 1.0
44.          elif wavelength >= 490 and wavelength <= 510:
45.              R = 0.0
46.              G = 1.0
47.              B = (-(wavelength - 510) / (510 - 490)) ** gamma
48.          elif wavelength >= 510 and wavelength <= 580:
49.              R = ((wavelength - 510) / (580 - 510)) ** gamma
50.              G = 1.0
51.              B = 0.0
52.          elif wavelength >= 580 and wavelength <= 645:
53.              R = 1.0
54.              G = (-(wavelength - 645) / (645 - 580)) ** gamma
55.              B = 0.0
56.          elif wavelength >= 645 and wavelength <= 750:
57.              attenuation = 0.3 + 0.7 * (750 - wavelength) / (750 - 645)
58.              R = (1.0 * attenuation) ** gamma
59.              G = 0.0
60.              B = 0.0
61.          else:
62.              R = 0.0
63.              G = 0.0
64.              B = 0.0
65.          return (R,G,B,A)
66.
```

# Index

Pages in *italics* refer to figures.

For Product Safety Concerns and Information please contact our EU
representative  GPSR@taylorandfrancis.com
Taylor & Francis Verlag GmbH, Kaufingerstraße 24, 80331 München, Germany